Formula One: The Grand Prix of Legends

A Chronological Journey through the World's Premier Motorsport

Etienne Psaila

Formula One: The Grand Prix of Legends

Copyright © 2023 by Etienne Psaila. All rights reserved.

First Edition: **November 2023**

No part of this publication may be reproduced, distributed, or transmitted in any form or by any means, including photocopying, recording, or other electronic or mechanical methods, without the prior written permission of the publisher, except in the case of brief quotations embodied in critical reviews and certain other non-commercial uses permitted by copyright law.

Cover design by Etienne Psaila

Interior layout by Etienne Psaila

Introduction

In the world of motorsport, few spectacles match the adrenaline and allure of Formula 1. A symphony of speed, skill, and innovation, Formula 1 has captivated the hearts of millions, carving its legacy on the tarmac of history. This book, "Speed and Glory: The Epic Saga of Formula 1," is a tribute to this magnificent sport, an exploration of its rich history and dynamic present.

From the roar of engines to the screech of tires, the journey of Formula 1 is not just about machines and regulations, but about the people who turned a racing competition into a global phenomenon. In these pages, we delve into the stories of legendary drivers, whose courage and charisma transcended the cockpit. Names like Juan Manuel Fangio, Ayrton Senna, and Lewis Hamilton are not merely participants; they are the heartbeats of this sport, exemplifying human resilience, skill, and the relentless pursuit of excellence.

But what of the cars themselves, these marvels of engineering that have evolved from mere fast cars to technological wonders? We explore the evolution of Formula 1 cars, tracing their transformation through the decades, each era bringing new challenges and innovations. From the sleek designs of the early years to the aerodynamically advanced machines of today, every car tells a story of ingenuity and the relentless pursuit of speed.

Behind every great driver and every powerful machine are the constructors and teams, the unsung heroes of Formula 1. This book sheds light on the pivotal role of teams like Ferrari, McLaren, and Mercedes, and how their rivalries and collaborations have shaped the sport. It's a world where strategy, technology, and human brilliance come together in a quest for racing supremacy.

Lastly, we cannot overlook the backbone of Formula 1 – the regulations that govern it. These rules have continually evolved, ensuring safety, fairness, and competition. We examine how these regulations have transformed over the years, shaping the sport into what it is today.

"Speed and Glory: The Epic Saga of Formula 1" is more than just a recounting of races and champions. It is a journey through time, capturing the spirit of innovation and the relentless human spirit driving one of the most exhilarating sports in the world. Welcome to the world of Formula 1, a world where every second counts and glory awaits at the finish line.

Chapter 1:

The Iconic Drivers

Juan Manuel Fangio (1950s)

Juan Manuel Fangio, often referred to as "El Maestro" (The Master), is one of the most revered figures in the history of Formula 1 racing. Born on June 24, 1911, in Balcarce, Argentina, Fangio's journey to becoming a five-time World Champion is a story of grit, determination, and unparalleled skill.

Early Life and Career Beginnings

- **Born in Argentina**: Fangio was born to Italian immigrant parents in a small town in Argentina. He started his career in motor racing not in Formula 1 but in long-distance races in South America, often on rough and dangerous terrain.

- **Racing in South America**: He gained fame in Argentina by winning the grueling "Gran Premio del Norte" in 1940, a race that stretched over 9,500 km across Argentina and Peru. This race took 13 days to complete and established Fangio's reputation as a resilient and talented driver.

Transition to European Racing and Formula 1

- **Move to Europe**: Fangio moved to Europe after World War II to pursue his racing career. His talents quickly caught the attention of European racing teams.

- **Debut in Formula 1**: He made his Formula 1 debut in the 1950 Monaco Grand Prix with the Alfa Romeo team. Although he was already 39, an age at which many drivers considered retirement, he quickly proved his worth.

Dominance in Formula 1

- **Multiple Championships**: Fangio won his first World Championship in 1951 with Alfa Romeo. He went on to win four more championships (1954, 1955, 1956, and 1957) with four different teams: Maserati, Mercedes-Benz, Ferrari, and again with Maserati.

- **Racing Style and Skills**: Fangio was known for his smooth driving style, precision, and ability to push cars to their limits without overstraining them. He was also a master of race tactics, often conserving his car's tires and brakes in the early stages of a race to have a better chance against his opponents towards the end.

Legendary Races and Records

- **German Grand Prix 1957**: Perhaps his most famous race was the 1957 German Grand Prix at the Nürburgring, where he made a remarkable comeback. After a pit stop that put him well behind the leaders, Fangio chased down and passed much younger drivers, setting multiple lap records in the process.

- **Records**: For many years, Fangio held the record for the most World Championships in Formula 1, a record only broken decades later by Michael Schumacher.

Retirement and Legacy

- **Retirement**: Fangio retired from Formula 1 in 1958, leaving behind an unmatched legacy. His record of winning five World Championships stood for 46 years until Michael Schumacher won his sixth title in 2003.

- **Legacy**: His legacy extends beyond his racing achievements. He was universally respected for his sportsmanship, humility, and gentlemanly conduct, both on and off the track. He was a role model for generations of racing drivers and remains a towering figure in the history of the sport.

Juan Manuel Fangio's story is not just one of triumph in the face of competition but also of the evolution of Formula 1 racing. His life and career symbolize the passion, skill, and dedication that define the essence of motorsport. His influence on the sport, the way he conducted himself, and the standards he set, continue to inspire racers and fans alike.

Stirling Moss (1950s-1960s)

Sir Stirling Moss, often hailed as "the greatest driver never to win the World Championship," is a legendary figure in the world of Formula 1 and motorsport. Born on September 17, 1929, in London, England, Moss's racing career is celebrated for his extraordinary skill and sportsmanship, as well as his versatility across different types of motorsports.

Early Life and Introduction to Racing

- **Born into a Racing Family**: Moss was born into a family with a motorsport background. His father was an amateur racer and his mother was a skilled driver as well.

- **Early Racing Career**: He began racing at a young age, competing in local events. His talent was evident from the start, as he quickly made a name for himself in the British racing scene.

Rising Through the Ranks

- **Success in Various Motorsport Disciplines**: Before entering Formula 1, Moss proved his versatility by competing successfully in rallies, hill climbs, and endurance races, showcasing his wide-ranging driving skills.

- **Joining Formula 1**: Moss entered Formula 1 in 1951. Although he drove for several teams throughout his career, his most successful period was with Mercedes-Benz, alongside Juan Manuel Fangio.

Near Misses and "The Best Without a Title"

- **Close Calls in Championships**: Despite being one of the most skilled drivers of his era, Moss never won a World Championship. He finished as runner-up in the drivers' championship four times and third on three occasions.

- **Notable Victories**: Moss's Formula 1 career included 16 Grand Prix victories. One of his most famous wins was the 1955 British Grand Prix, where he became the first British driver to win his home Grand Prix.

Record-Setting and Memorable Races

- **Mille Miglia 1955**: Among his most extraordinary achievements was winning the 1955 Mille Miglia, a grueling 1,000-mile race across Italy. He completed the race in just over 10 hours, setting an average speed record that stands to this day.

- **Versatility Across Races**: Moss's ability to compete at the highest level in different types of racing, including sports cars and endurance races, set him apart from many of his contemporaries.

Sportsmanship and Personality

- **Sportsmanship**: Moss was known for his unwavering sportsmanship. In the 1958 Portuguese Grand Prix, he defended rival Mike Hawthorn, who was facing disqualification; Hawthorn's points from that race ultimately helped him beat Moss to the championship.

- **Charismatic and Popular**: Beyond his racing achievements, Moss was a charismatic figure, known for his gentlemanly conduct and popularity among fans and fellow drivers alike.

Career End and Legacy

- **Career-Ending Crash**: Moss's career was abruptly ended by a crash at Goodwood in 1962, which left him in a coma for a month and partially paralyzed for six months.

- **Legacy**: Despite never winning a World Championship, Moss remains one of the most respected figures in motorsport history. His skill, sportsmanship, and versatility have made him a symbol of the golden era of racing.

Sir Stirling Moss's story is one of extraordinary talent, character, and determination. His achievements in various forms of motorsport and his approach to competition have left an indelible mark on the world of racing, making him a true icon of the sport.

Jim Clark (1960s)

Jim Clark, considered by many as one of the greatest Formula 1 drivers of all time, was known for his exceptional talent and modest personality. Born on March 4, 1936, in Kilmany, Scotland, Clark's career in Formula 1 during the 1960s was marked by dominance, innovation, and tragedy.

Early Life and Racing Beginnings

- **Scottish Farmer Turned Racer**: Clark grew up on a farm in Scotland. He started his racing career in local road rallies and hill climbs, driving a variety of cars.

- **Initial Success**: His natural talent quickly shone through, and he soon caught the attention of the motorsport community.

Joining Formula 1

- **Debut with Lotus**: Clark made his Formula 1 debut in 1960 with Lotus, a team he would remain with throughout his F1 career. He was closely associated with team founder and design innovator, Colin Chapman.

- **First Victory**: Clark won his first Grand Prix in Belgium in 1962, showcasing his extraordinary driving skills.

Period of Dominance

- **World Championships**: Clark won the Formula 1 World Championship twice, in 1963 and 1965. His first championship in 1963 was particularly dominant; he won seven out of ten races.

- **1965 Indianapolis 500**: In addition to his F1 success, Clark also competed in the Indianapolis 500, winning it in 1965. This victory made him the first non-American to win since 1916 and the only driver to win both the Indy 500 and the Formula 1 World Championship in the same year.

Skill and Driving Style

- **Exceptional Talent**: Clark was known for his smooth driving style, precision, and ability to extract the maximum from his car. He was often faster than his teammates by significant margins.

- **Versatility**: Besides his F1 and Indy 500 successes, Clark was competitive in various other motorsport disciplines, including touring car and sports car racing.

Tragedy and Legacy

- **Tragic Death**: Jim Clark's life and career were tragically cut short at the age of 32. He died in a Formula 2 crash at Hockenheim in Germany in 1968.

- **Legacy**: Despite his premature death, Clark left an indelible mark on Formula 1 and motorsport. He is remembered for his humble personality, sportsmanship, and his natural, seemingly effortless driving ability.

- **Respect and Remembrance**: Clark was deeply respected by his peers and loved by fans. His death led to increased emphasis on safety in motorsport, a cause that continued to gain momentum in the following decades.

Jim Clark's story is one of extraordinary natural talent and humility, making him a revered figure in the history of motorsport. His achievements during the golden era of Formula 1 continue to inspire and resonate with racing fans and drivers alike.

Graham Hill (1960s)

Graham Hill, a prominent figure in the world of Formula 1 during the 1960s, is remembered not only for his racing achievements but also for his charismatic personality and distinguished appearance. Born on February 15, 1929, in Hampstead, London, Hill's career in Formula 1 was characterized by success, versatility, and resilience.

Early Life and Entry into Racing

- **Late Start in Motorsport**: Unlike many of his contemporaries, Hill did not start racing until his mid-twenties. Before his racing career, he served in the Royal Navy and pursued various jobs.

- **Initial Struggles**: Hill's entry into professional racing wasn't easy. He struggled initially, working as a mechanic and undertaking minor driving roles before getting his break in Formula 1.

Formula 1 Career

- **Debut and Early Years**: Graham Hill made his Formula 1 debut in 1958 with Team Lotus. His early years were challenging, but he persevered, gradually improving his skills and reputation.

- **World Championships**: Hill won his first World Championship in 1962 with BRM (British Racing Motors). He won his second title in 1968 with Lotus, showcasing his exceptional driving skills and determination.

Versatility and the Triple Crown

- **Remarkable Versatility**: Hill was known for his versatility across different forms of motorsport. He is the only person to have won the Triple Crown of Motorsport: the Monaco Grand Prix (which he won five times), the Indianapolis 500 (in 1966), and the Le Mans 24 Hours race (in 1972).

- **Endurance Racing**: Besides his Formula 1 achievements, Hill excelled in sports car racing, including his victory at Le Mans.

Personality and Legacy

- **Charisma and Appearance**: Hill was known for his dapper appearance, often sporting a neatly trimmed mustache. He was charismatic, witty, and popular among fans and within the racing community.

- **Tragic End and Legacy**: Graham Hill's life tragically ended in a plane crash in 1975. His death was a significant loss to the motorsport world.

- **Family Legacy**: His legacy in motorsport continued through his son, Damon Hill, who also became a Formula 1 World Champion in 1996.

Contribution to Motorsport

- **Team Owner**: In the latter part of his career, Hill formed his own racing team, Embassy Hill, through which he contributed to the development of the sport beyond his role as a driver.

- **Mentorship and Influence**: Hill was known for mentoring younger drivers and was respected for his experience and knowledge of the sport.

Graham Hill's story in Formula 1 is one of determination, versatility, and charisma. His unique achievement of winning the Triple Crown, along with his success in Formula 1, makes him one of the most celebrated and respected figures in the history of motorsport. His personality, coupled with his sporting achievements, left a lasting impression on the world of racing.

Jackie Stewart (1960s-1970s)

Sir Jackie Stewart, also known as "The Flying Scot," is one of the most influential figures in the history of Formula 1, both for his achievements on the track and his tireless advocacy for safety in the sport. Born on June 11, 1939, in Milton, West Dunbartonshire, Scotland, Stewart's Formula 1 career spanned the late 1960s and early 1970s, a period known for its high risk and minimal safety standards.

Early Career and Entry into Formula 1

- **Early Struggles with Dyslexia:** Stewart struggled with undiagnosed dyslexia in his early years, which affected his schooling. He initially pursued a career in clay pigeon shooting, showing considerable talent.

- **Entry into Racing:** Stewart began his racing career in the early 1960s, quickly moving up from driving touring cars to Formula 3 and then to Formula 1 in 1965 with BRM (British Racing Motors).

Formula 1 Success

- **First Victory**: His first Formula 1 victory came in the 1965 Italian Grand Prix. He was known for his smooth, controlled driving style and his ability to read the race and conditions effectively.

- **World Championships**: Stewart won three Formula 1 World Championships in 1969, 1971, and 1973. His driving skill, combined with his intelligence and tactical acumen, made him a formidable competitor.

Advocacy for Safety

- **Passion for Safety**: After witnessing several fatal accidents, including the death of his friend and fellow driver, Jochen Rindt, Stewart became a vocal advocate for improving safety standards in Formula 1.

- **Implementing Changes**: He campaigned for better medical facilities, safer track conditions, and the use of seat belts and full-face helmets. His efforts led to significant changes in safety regulations, undoubtedly saving many lives.

Retirement and Post-Racing Career

- **Retirement from Racing**: Stewart retired from Formula 1 in 1973, at the peak of his career, partly due to the dangers involved in the sport at the time.

- **Continued Influence**: After retiring, he remained involved in Formula 1, including team management, commentary, and continuing his safety advocacy. He also became a successful businessman and an ambassador for the sport.

Legacy and Recognition

- **Respect and Admiration**: Stewart is widely respected for his contributions to Formula 1, both as a supremely talented driver and as a pioneer in the field of safety.

- **Honors and Awards**: He was knighted in 2001 for his services to motorsport. His legacy in the sport is not just measured in races won but also in the countless lives he helped protect through his safety efforts.

Sir Jackie Stewart's story in Formula 1 is a compelling tale of triumph, tragedy, and transformation. His remarkable driving career, coupled with his passionate advocacy for safety, has left an indelible mark on the sport, making him a true icon in the world of motorsports.

Niki Lauda (1970s-1980s)

Niki Lauda, an Austrian racing legend, is renowned for his exceptional driving skills, remarkable comeback story, and contributions to Formula 1 both on and off the track. Born on February 22, 1949, in Vienna, Austria, Lauda's career in Formula 1 during the 1970s and 1980s was marked by both triumph and adversity.

Early Career and Entry into Formula 1

- **Struggle Against Family Expectations**: Lauda came from a wealthy family, but his ambitions to be a race car driver were not supported by his family. He pursued his passion regardless, taking bank loans to fund his early racing endeavors.

- **Entry into Formula 1**: Lauda made his Formula 1 debut in 1971. His talent was evident, but initial success was limited. His breakthrough came when he joined Ferrari in 1974.

Rise to Prominence

- **First World Championship**: Lauda won his first World Championship with Ferrari in 1975, showcasing his meticulous approach to racing and technical understanding of the cars.

- **Dominant 1976 Season**: The 1976 season began with Lauda dominating, winning four of the first six races.

Near-Fatal Accident and Remarkable Comeback

- **Horrific Accident at Nürburgring**: Lauda's life and career dramatically changed at the 1976 German Grand Prix at the Nürburgring, where he suffered a near-fatal accident, leaving him with severe burns and lung damage.

- **Incredible Recovery**: In one of the most remarkable comebacks in sports history, Lauda returned to racing just six weeks after his accident, still in pain and bandaged. His determination and bravery were awe-inspiring.

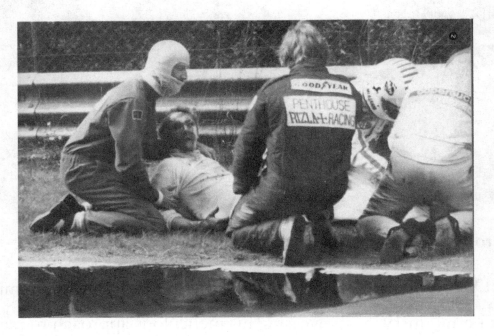

Further Championships and Rivalries

- **1977 World Championship**: Despite the accident, Lauda won his second World Championship in 1977. His ability to overcome adversity and compete at the highest level was a testament to his resilience and skill.

- **Rivalry with James Hunt**: Lauda's rivalry with British driver James Hunt, particularly during the 1976 season, is one of the most famous in Formula 1 history, characterized by mutual respect and intense competition.

Later Career and Retirement

- **Third World Championship**: After a brief retirement, Lauda returned to Formula 1 in 1982 with McLaren and went on to win his third World Championship in 1984.

- **Retirement and Beyond**: Lauda retired for good in 1985. He then pursued various roles, including becoming a successful airline entrepreneur and later serving as a team manager and advisor in Formula

Legacy

- **Contribution to Motorsport**: Lauda is remembered not only for his racing achievements but also for his contributions to team management and driver safety.

- **Respected Figure**: His analytical approach to racing, combined with his straightforwardness and resilience, made him a highly respected figure in the world of motorsports.

Niki Lauda's story is one of extraordinary determination, skill, and courage. His legacy in Formula 1 extends beyond his championships, embodying the spirit of resilience and the relentless pursuit of excellence. His life and career continue to inspire and resonate within the racing community and beyond.

Alain Prost (1980s-1990s)

Alain Prost, nicknamed "The Professor" for his intellectual approach to racing, is a renowned figure in Formula 1, known for his tactical brilliance and smooth driving style. Born on February 24, 1955, in Lorette, Loire, France, Prost's career spanned the 1980s and early 1990s, a period marked by intense rivalries and significant technological advancements in the sport.

Early Career and Entry into Formula 1

- **Karting and Early Racing**: Prost began his career in karting and quickly moved through the ranks in various junior formulas, showcasing his talent and determination.

- **Formula 1 Debut**: He made his Formula 1 debut in 1980 with McLaren, immediately making an impression with his speed and racing intelligence.

Rise to Prominence

- **First Wins and Championship Challenges**: Prost claimed his first Formula 1 victory in 1981 with Renault. His early years in the sport were marked by consistent performances and championship challenges, setting the stage for his future success.

World Championships

- **McLaren and Championships**: Prost won his first World Championship in 1985 with McLaren, a team with which he would become closely associated. He went on to win three more championships (1986, 1989, and 1993), making him one of the most successful drivers in the history of the sport.

- **Rivalry with Ayrton Senna**: Perhaps the most defining aspect of Prost's career was his rivalry with teammate Ayrton Senna. Their battles on the track were fierce and often controversial, but they pushed the limits of the sport and are remembered as some of the greatest moments in Formula 1 history.

Driving Style and Approach

- **Tactical Intelligence**: Prost was known for his strategic approach to racing, often thinking several laps ahead and conserving his car and tires for critical moments in the race.

- **Smooth Driving Style**: His driving style was smooth and efficient, which allowed him to excel, especially in races that required careful management of the car.

Later Career and Retirement

- **Brief Period with Ferrari**: After a stint with Ferrari, where he won several races but faced challenges, Prost took a sabbatical from Formula 1 in 1992.

- **Return and Final Championship with Williams**: He returned in 1993 with Williams, winning his fourth and final World Championship before retiring from the sport.

Post-Racing Career and Legacy

- **Team Management and Commentary**: After retiring, Prost remained active in the sport, including a stint as a team owner and serving as a commentator and advisor.

- **Influence on the Sport**: Prost's influence on Formula 1 goes beyond his race wins and championships. His intellectual approach to racing and his rivalry with Senna helped define an era in the sport.

Alain Prost's story in Formula 1 is one of precision, intelligence, and strategic mastery. His legacy as one of the sport's greatest drivers is secured not just by his multiple championships but also by the way he revolutionized the approach to racing, combining technical understanding with a tactical mindset. His career, marked by intense rivalries and iconic moments, continues to be a benchmark in the annals of motorsport history.

Ayrton Senna (1980s-1990s)

Ayrton Senna, widely regarded as one of the greatest Formula 1 drivers of all time, is remembered for his incredible skill, particularly in wet conditions, and his intense dedication to the sport. Born on March 21, 1960, in São Paulo, Brazil, Senna's career in the 1980s and early 1990s was marked by remarkable victories, fierce rivalries, and an enduring legacy.

Early Life and Racing Beginnings

- **Karting Prodigy**: Senna's passion for racing began in childhood, with karting. He showed remarkable talent early on, winning numerous karting titles.

- **Moving Through the Ranks**: He progressed through various junior motorsport categories, showcasing his exceptional talent and intense dedication.

Entry into Formula 1

- **Formula 1 Debut**: Senna made his Formula 1 debut in 1984 with Toleman. Despite driving for a relatively modest team, his talent was undeniable, particularly highlighted during the Monaco Grand Prix, where he delivered a stunning performance in the rain.

- **Move to Lotus**: In 1985, Senna joined Lotus, where he achieved his first Grand Prix win. He won multiple races with Lotus, often outperforming the capabilities of his car.

Dominance and Rivalries

- **McLaren Years**: Senna's move to McLaren in 1988 marked the beginning of his most successful period. He won his first World Championship that year, showcasing his extraordinary skill.

- **Rivalry with Alain Prost**: At McLaren, Senna developed a fierce rivalry with teammate Alain Prost. Their battles on and off the track became one of the most famous rivalries in sports history, marked by both intense competition and profound respect.

- **Three World Championships**: Senna won three World Championships (1988, 1990, 1991), becoming known for his exceptional performances, especially in challenging conditions.

Driving Style and Dedication

- **Master in Wet Conditions**: Senna had an uncanny ability to drive in wet conditions, where his skill and intuition shone brightly. His performances in the rain are considered some of the best in Formula 1 history.

- **Relentless Pursuit of Perfection**: Known for his intense focus and determination, Senna continuously pushed himself and his team to the limits, seeking every possible advantage.

Tragic Death and Legacy

- **Death at Imola**: Tragically, Senna's life and career were cut short on May 1, 1994, when he suffered a fatal crash at the San Marino Grand Prix in Imola, Italy. His death led to profound changes in the sport's safety standards.

- **Legacy**: Ayrton Senna's legacy extends far beyond his race wins and championships. He is remembered for his passionate and spiritual approach to racing, his humanitarian efforts, and his profound impact on Formula 1. Senna's dedication, charisma, and driving prowess have made him an enduring icon in the sport.

Senna's story in Formula 1 is one of brilliance, intensity, and emotion. His commitment to excellence, combined with his human qualities, continues to inspire drivers and fans worldwide, making him a timeless figure in the annals of motorsport history.

Michael Schumacher (1990s-2000s)

Michael Schumacher, a German racing driver, is widely regarded as one of the greatest Formula 1 drivers in history. His career, spanning the 1990s and 2000s, was characterized by record-breaking achievements, immense skill, and a profound impact on the sport.

Early Career and Entry into Formula 1

- **Karting Beginnings**: Born on January 3, 1969, in Hürth, Germany, Schumacher's passion for racing began with karting at a young age, where he quickly made a name for himself.

- **Climbing the Ranks**: He progressed through various motorsport categories, including Formula Ford and Formula 3, showcasing his talent and determination.

Rise to Prominence in Formula 1

- **Formula 1 Debut**: Schumacher made his Formula 1 debut in 1991 with the Jordan team. His performance immediately caught the attention of the racing world, and he was quickly signed by Benetton.

- **First Championships with Benetton**: He won his first World Championship in 1994, followed by another in 1995. These victories were marked by both brilliance and controversy due to various incidents and the intense competition of the era.

Dominance at Ferrari

- **Move to Ferrari**: In 1996, Schumacher joined Ferrari, a team that had been struggling to achieve success. His arrival marked the beginning of one of the most remarkable periods in the history of Formula 1.

- **Creating a Legacy**: At Ferrari, Schumacher, along with team principal Jean Todt, technical director Ross Brawn, and chief designer Rory Byrne, turned the team into a dominant force, winning five consecutive World Championships from 2000 to 2004.

Driving Style and Work Ethic

- **Exceptional Skill**: Schumacher was known for his exceptional driving skill, particularly in wet conditions and his ability to extract the best performance from his car.

- **Work Ethic and Dedication**: He was also renowned for his work ethic, attention to detail, and physical fitness, which set new standards in Formula 1.

Records and Achievements

- **Record-Breaking Success**: Schumacher set numerous records during his career, including seven World Championships (two with Benetton and five with Ferrari), 91 Grand Prix victories, and the most fastest laps.

- **Influence on the Sport**: Beyond his racing achievements, Schumacher's approach to testing, fitness, and teamwork influenced how Formula 1 teams operate today.

Retirement and Comeback

- **First Retirement**: Schumacher initially retired in 2006 but remained involved with Ferrari in various capacities.

- **Return with Mercedes**: He made a comeback to Formula 1 with Mercedes in 2010, racing for three seasons before retiring again in 2012.

Legacy

- **Respected Figure in Motorsport**: Schumacher's impact on Formula 1 is immense, not just in terms of his records and victories, but also in how he transformed team dynamics and driver preparation.

- **Humanitarian Efforts**: Off the track, he was known for his humanitarian and charity work.

Michael Schumacher's story in Formula 1 is one of extraordinary success, relentless pursuit of perfection, and a lasting impact on the sport. His achievements, dedication, and the standards he set continue to inspire and influence the world of motorsport.

Fernando Alonso (2000s-Present)

Fernando Alonso, a Spanish racing driver, is celebrated for his exceptional talent, competitive spirit, and substantial impact on Formula 1. His career, initially spanning the early 2000s to the 2010s and continuing presently, features notable successes, adaptability, and a fiercely competitive demeanor.

Early Career and Entry into Formula 1

- **Karting Beginnings**: Born on July 29, 1981, in Oviedo, Spain, Alonso's racing career started in karting, where he clinched multiple championships, demonstrating early signs of exceptional driving skills.
- **Rising Through the Ranks**: Alonso rapidly ascended through junior motorsport ranks, impressing with his speed and race intelligence.

Breakthrough in Formula 1

- **Formula 1 Debut**: Alonso debuted in Formula 1 in 2001 with Minardi, where, despite limited resources, he displayed significant potential.

- **Joining Renault**: His move to Renault in 2003 was a pivotal moment, as he became the then-youngest Grand Prix winner at the 2003 Hungarian Grand Prix.

World Championships and Rivalries

- **Youngest World Champion**: In 2005, Alonso secured his first World Championship with Renault, becoming the youngest champion in Formula 1 history at that time. He secured another championship in 2006, cementing his status as a top-tier driver.

- **Rivalry with Schumacher**: His championship years were marked by a notable rivalry with Michael Schumacher, representing a shift in Formula 1's competitive landscape.

Tenure with Various Teams

- **Controversial Season at McLaren**: In 2007, Alonso joined McLaren, experiencing a competitive yet tumultuous season, ultimately finishing third in the championship.

- **Return to Renault**: Alonso went back to Renault for two seasons, maintaining competitiveness in a less dominant car.

- **Ferrari Era**: In 2010, Alonso began racing for Ferrari. Despite several close championship bids, he finished runner-up three times during his stint with the team.

Driving Style and Prowess

- **Adaptability and Versatility**: Alonso is celebrated for his adaptability to diverse cars and conditions, often maximizing performance from less competitive vehicles.

- **Masterful Racecraft**: His exceptional racecraft, overtaking capabilities, and unwavering determination are highly regarded in the sport.

Continued Career and Legacy

- **Second McLaren Period and Sabbatical**: Alonso's second tenure with McLaren was challenging due to uncompetitive cars. He took a break from Formula 1 after the 2018 season.

- **Endurance Racing and Return to F1**: During his sabbatical, Alonso ventured into other motorsport disciplines, including winning the Le Mans 24 Hours and participating in the Indy 500. He made a Formula 1 comeback in 2021 with Alpine.

- **Current Racing with Aston Martin**: As of now, Alonso continues his Formula 1 journey, racing with Aston Martin, demonstrating his enduring skill and passion for the sport.

- **Enduring Influence**: Alonso is revered for his driving skills, competitive nature, and capability to maintain high performance across various motorsport forms.

Fernando Alonso's narrative in Formula 1 is a testament to resilience, versatility, and raw talent. His journey through multiple teams, coupled with his triumphs and relentless competitive spirit, has established him as one of the most respected and adaptable drivers in the sport's history.

Lewis Hamilton (2000s-present)

Lewis Hamilton, a British racing driver, is one of the most successful and influential figures in the history of Formula 1. His career, which began in the mid-2000s and continues to this day, is distinguished by record-breaking achievements, a groundbreaking impact on the sport, and a strong commitment to social and environmental causes.

Early Life and Racing Beginnings

- **Karting Prodigy**: Born on January 7, 1985, in Stevenage, Hertfordshire, England, Hamilton's journey into racing started with karting at a young age, where he quickly showed extraordinary talent.

- **Support from McLaren**: His potential was recognized early on by McLaren, which provided support and guidance through his junior racing years.

Entry into Formula 1

- **Formula 1 Debut**: Hamilton made a sensational debut in Formula 1 in 2007 with McLaren, almost winning the World Championship in his rookie year.

- **First World Championship**: He won his first World Championship in 2008, becoming the then-youngest World Champion in Formula 1 history.

Record-Breaking Success at Mercedes

- **Move to Mercedes**: In 2013, Hamilton made a pivotal move to Mercedes. This partnership led to one of the most dominant eras in Formula 1 history.

- **Multiple World Championships**: At Mercedes, Hamilton has won numerous World Championships, breaking and setting several records, including the most race wins, pole positions, and podium finishes.

Driving Style and Skills

- **Exceptional Talent**: Hamilton is known for his exceptional driving skills, particularly in wet conditions and his ability to perform consistently at a high level.

- **Racecraft and Adaptability**: His racecraft, strategic intelligence, and adaptability are considered among the best in the sport's history.

Off-Track Influence

- **Advocacy for Diversity and Inclusion**: Beyond his racing achievements, Hamilton is a vocal advocate for diversity and inclusion in motorsport. He has actively worked to promote opportunities for underrepresented groups in the sport.

- **Environmental and Social Activism**: He is also known for his environmental activism and commitment to social issues, using his platform to raise awareness and drive positive change.

Legacy and Impact

- **One of the Greatest Drivers**: Hamilton's ongoing career has already secured his status as one of the greatest drivers in the history of Formula 1.

- **Role Model**: His impact extends beyond the track, as he has become a role model for aspiring drivers and an influential figure in global discussions on social and environmental matters.

Lewis Hamilton's story in Formula 1 is not just one of unparalleled success and record-breaking achievements but also of breaking barriers and inspiring change. His legacy, both on and off the track, continues to evolve, marking him as a unique and transformative figure in the world of sports.

Sebastian Vettel (2000s-2022)

Sebastian Vettel, a German racing driver, stands out as one of the most accomplished and esteemed figures in the history of Formula 1. Spanning from the mid-2000s and culminating in his retirement, his career is marked by significant achievements, strategic mastery, and a profound influence on the sport.

Early Career and Entry into Formula 1

- **Karting Beginnings**: Born on July 3, 1987, in Heppenheim, Germany, Vettel's racing journey started in karting, showcasing exceptional potential early on.

- **Climbing the Ranks**: He swiftly moved through junior motorsport series, displaying remarkable skill and promise in each category.

Formula 1 Breakthrough

- **Formula 1 Debut**: Vettel's Formula 1 debut came in 2007 with BMW Sauber as a replacement driver, later joining Toro Rosso. His talent quickly became evident.

- **First Formula 1 Victory**: He claimed his first Grand Prix victory in 2008 at the Italian Grand Prix with Toro Rosso, becoming the youngest driver at that time to win a Formula 1 race.

Dominance with Red Bull Racing

- **Joining Red Bull Racing**: In 2009, Vettel moved to Red Bull Racing, initiating a period of significant success for both him and the team.

- **World Championships**: From 2010 to 2013, Vettel won four consecutive World Championships, demonstrating his extraordinary driving ability and strategic acumen.

- **Setting Records**: His tenure at Red Bull included setting multiple records, such as the most consecutive race wins and becoming the youngest World Champion.

Subsequent Career and Retirement

- **Ferrari Tenure**: Vettel joined Ferrari in 2015, aspiring to replicate Michael Schumacher's success. Despite several victories and title challenges, he did not secure a World Championship with the team.

- **Transition to Aston Martin**: After his Ferrari stint, Vettel joined Aston Martin in 2021, bringing his vast experience to the team and continuing to compete at a high level.

- **Retirement Announcement**: Vettel announced his retirement from Formula 1 at the end of the 2022 season, concluding a storied and impactful career.

Driving Skills and Style

- **Tactical Brilliance**: Known for his tactical intelligence, exceptional qualifying speed, and proficiency in leading races, Vettel's driving style has been a key to his success.

- **Versatility**: His adaptability to various cars and conditions highlights his versatile driving capabilities.

Off-Track Contributions

- **Personality and Advocacy**: Vettel is renowned for his congenial personality, sportsmanship, and active advocacy on environmental and social issues.

- **Mentoring Young Drivers**: As a seasoned driver, he played a mentorship role, imparting knowledge and guidance to emerging talents.

Legacy and Impact

- **One of Formula 1's Greats**: Vettel's multiple championships and race victories solidify his status as one of the greats in Formula 1, celebrated not only for his on-track successes but also for his contributions to the sport's culture.

- **Respected and Influential Figure**: His respectful approach, commitment to various causes, and enduring competitiveness have earned him widespread admiration in the motorsport community and beyond.

Sebastian Vettel's journey in Formula 1 is a narrative of remarkable triumph, consistent performance, and influential presence, culminating in a retirement that marks the end of an era. His legacy extends beyond his race wins and championships, encompassing his role as a sportsman and advocate for positive change both within the sport and in wider societal contexts.

Max Verstappen (2010s-present)

Max Verstappen, a Belgian-Dutch racing driver, has rapidly become one of the standout figures in Formula 1, known for his aggressive driving style, remarkable talent, and significant achievements. His career, starting in the mid-2010s and continuing presently, has been marked by record-breaking performances and a fierce competitive spirit.

Early Career and Formula 1 Entry

- **Karting and Early Racing**: Born on September 30, 1997, in Hasselt, Belgium, Verstappen's journey in motorsports began in karting. He quickly made a name for himself with his exceptional talent and determination.

- **Rapid Ascent**: Verstappen's progression through the racing ranks was meteoric, swiftly moving from karting to Formula 3 and then to Formula

Formula 1 Breakthrough

- **Youngest Ever Formula 1 Driver**: Verstappen made his Formula 1 debut with Toro Rosso in 2015 at the age of 17, becoming the youngest ever driver to compete in Formula 1.

- **First Formula 1 Win**: He claimed his first Grand Prix victory with Red Bull Racing in 2016 at the Spanish Grand Prix, immediately after being promoted from Toro Rosso, becoming the youngest driver ever to win a Formula 1 race.

Rising Star and World Championships

- **Consistent Competitor**: Over the following years, Verstappen established himself as a consistent top competitor, known for his bold overtaking maneuvers and aggressive racing style.

- **2021 World Champion**: Verstappen won his first World Championship in 2021 in a dramatic and controversial final race, ending the season-long intense rivalry with Lewis Hamilton.

- **Dominance in 2023**: In 2023, Verstappen dominated the World Championship, breaking various records and showcasing his skill and maturity as a driver. His performance throughout the season was marked by strategic brilliance and exceptional driving.

Driving Style and Skills

- **Aggressive and Precise**: Verstappen is known for his aggressive yet precise driving style, remarkable racecraft, and ability to perform under pressure.

- **Adaptability and Resilience**: His adaptability to different race conditions and resilience in challenging situations have been key to his success.

Off-Track Persona

- **Confident and Focused**: Off the track, Verstappen is known for his confidence and focus. He maintains a relatively private life, with a clear dedication to his racing career.
- **Growing Influence**: As one of the younger drivers on the grid, Verstappen's influence in the sport is growing, appealing to a new generation of Formula 1 fans.

Legacy and Impact

- **Redefining Records**: Verstappen's career, particularly his 2023 World Championship domination, is redefining what is possible for younger drivers in the sport.

- **Inspiring a New Era**: His success and competitive nature are inspiring a new era in Formula 1, marked by youthful energy and a challenge to the established order.

Max Verstappen's story in Formula 1 is one of rapid ascent, remarkable talent, and record-breaking achievements. His aggressive driving, combined with strategic acumen, continues to make him one of the most exciting drivers to watch on the grid. His impact on the sport extends beyond his race wins, influencing the next generation of Formula 1 racing and fans alike.

Chapter 2:

The Iconic Cars

Mercedes W196 (1954-1955)

The Mercedes W196, which raced in the 1954 and 1955 Formula 1 seasons, is a car steeped in history and innovation, remembered as much for its technological advancements as for its success on the track.

Background and Development

- **Return of Mercedes-Benz to Racing**: The W196 marked the return of Mercedes-Benz to Grand Prix racing after World War II. The company aimed to re-establish its dominance in motorsport, a status it had enjoyed in the pre-war era.

- **Technical Innovation**: The car was developed with advanced technology for its time, including a fuel-injected straight-8 engine, which was a first in Formula 1. It also featured a desmodromic valve actuation system for improved engine performance.

Design and Features

- **Streamlined Body**: One of the most distinctive features of the W196 was its streamlined body. The car debuted with a fully enclosed, narrow body aimed at reducing aerodynamic drag, which was ideal for high-speed tracks.

- **Open-Wheel Version**: For tighter circuits where handling was more critical, Mercedes developed an open-wheel version of the W196, providing better visibility and maneuverability.

Racing Success

- **Debut and Immediate Impact**: The W196 made its debut at the 1954 French Grand Prix at Reims-Gueux. Juan Manuel Fangio, driving the W196, won the race, demonstrating the car's superiority.

- **Championships**: Fangio went on to win the World Championship in both 1954 and 1955 with the W196, dominating the competition. The car won 9 out of the 12 races it entered, showcasing its technical and performance edge.

Juan Manuel Fangio and Stirling Moss

- **Fangio's Dominance**: Juan Manuel Fangio was the primary driver for Mercedes during this period. His skill combined with the W196's advanced technology made for an unbeatable combination.

- **Stirling Moss's Contributions**: Stirling Moss also drove the W196, most notably winning the 1955 British Grand Prix, making him the first British driver to win his home Grand Prix.

Legacy and Impact

- **Technological Influence**: The technological advancements of the W196 influenced Formula 1 car design for years to come. Its success demonstrated the importance of continuous innovation in the sport.

- **Retirement and Preservation**: Mercedes-Benz withdrew from motor racing at the end of the 1955 season following the Le Mans disaster. The W196's racing career was brief but left a lasting legacy. Today, the W196 is a prized and valuable collector's item, representing a pivotal era in Formula 1 history.

The Mercedes W196 is not just a car; it's a symbol of post-war resurgence, technological ambition, and racing excellence. Its story is a testament to the impact that innovative design and engineering can have in motorsport, setting new standards and pushing the boundaries of what was possible in its time.

Ferrari 156 'Sharknose' (1961)

The Ferrari 156, famously nicknamed the "Sharknose" due to its distinctive front-end design, is a legendary Formula 1 car that played a pivotal role in Ferrari's racing history. Its time in the spotlight was during the 1961 Formula 1 season.

Background and Development

- **Response to Rule Changes**: The development of the Ferrari 156 was in response to the new Formula 1 regulations introduced in 1961, which limited engine capacity to 1.5 liters. Ferrari's engineers, led by Carlo Chiti, designed the 156 to comply with these new rules.

- **Innovative Design**: The car featured a sleek, aerodynamic design, with the most notable feature being its uniquely shaped nose, resembling the snout of a shark, which gave the car its iconic nickname.

Technical Features

- **Engine and Performance**: The 156 was powered by a 1.5-liter V6 engine, known as the "Dino" after Enzo Ferrari's late son, Alfredo "Dino" Ferrari. The engine was notable for its use of a 120° angle, which helped lower the car's center of gravity and improve handling.

- **Chassis and Handling**: The car's chassis was relatively simple and lightweight, contributing to its agility and responsiveness on the track.

Racing Achievements

- **Dominant Season**: The 1961 Formula 1 season was dominated by the Ferrari 156. The car's combination of powerful engine performance and superior handling made it the car to beat.

- **Phil Hill's Championship**: American driver Phil Hill drove the 156 to win the 1961 World Championship, becoming the first American to achieve this feat. His victory was also Ferrari's first World Championship with a rear-engine car.

- **Monza Tragedy**: The 1961 season was also marked by tragedy. Wolfgang von Trips, driving a 156, was involved in a fatal accident at the Italian Grand Prix at Monza, which also claimed the lives of 15 spectators.

Legacy and Impact

- **Iconic Status**: The "Sharknose" Ferrari 156 is remembered as one of the most iconic and beautiful Formula 1 cars. Its innovative design and success on the track helped cement Ferrari's reputation as a leading force in Formula 1.

- **End of an Era**: The 156's success was short-lived, as the car struggled to compete effectively in the subsequent seasons. However, its impact on Ferrari's racing legacy and Formula 1 design was enduring.

- **Modern Interest**: Original 156s are extremely rare, as Ferrari was known for repurposing components from older cars. Replicas have been created, and the 156 "Sharknose" continues to be a highlight at historic racing events.

The Ferrari 156 "Sharknose" stands as a testament to Ferrari's innovative spirit and racing prowess in the early 1960s. Its distinctive design, technological advancements, and racing success have made it a beloved and unforgettable chapter in the annals of Formula 1 history.

Lotus 49 (1967-1970)

The Lotus 49, a Formula 1 car that competed from 1967 to 1970, is celebrated for its revolutionary design and significant impact on the sport. Designed by Colin Chapman and Maurice Philippe for Team Lotus, the Lotus 49 was a game-changer in Formula 1 engineering and aesthetics.

Background and Development

- **Innovative Concept**: The Lotus 49 was developed to accommodate the new Ford Cosworth DFV engine, which would become one of the most successful engines in Formula 1 history. Chapman's innovative approach integrated the engine as a stress-bearing structural component of the car, a novel idea at the time.

- **Collaboration with Ford**: The development of the Cosworth DFV engine was funded by Ford. This V8 engine was powerful yet lightweight, perfectly complementing the Lotus 49's design philosophy.

Design and Technical Features

- **Chassis Innovation**: The Lotus 49 was one of the first Formula 1 cars to use the engine as a load-bearing structural element, connected to the monocoque at the front and the suspension and gearbox at the rear. This design significantly reduced weight and increased rigidity.

- **Aerodynamic Efficiency**: While initially designed with minimal aerodynamic embellishments, the Lotus 49's sleek and straightforward design laid the groundwork for future aerodynamic exploration in Formula 1.

Racing Success

- **Debut and Immediate Impact**: The Lotus 49 made an immediate impact in its debut at the 1967 Dutch Grand Prix, with Jim Clark taking a remarkable victory.

- **World Championships**: The car won two World Championships: Graham Hill won in 1968, and Jochen Rindt posthumously in 1970, showcasing the car's competitiveness and advanced design.

- **Numerous Victories**: Throughout its racing life, the Lotus 49 claimed multiple Grand Prix victories, competing effectively against more technically advanced cars in its later years.

Legacy and Impact

- **Engineering Benchmark**: The Lotus 49 set a new standard in Formula 1 car design. Its integration of the engine as a structural component of the chassis became a fundamental principle in F1 car design.

- **Iconic Status**: The car is remembered not only for its technological innovations but also for its beautiful and sleek design. It remains one of the most visually iconic and influential Formula 1 cars.

- **Influence on Future Designs**: The design philosophies introduced by the Lotus 49 influenced the development of future racing cars, both within Lotus and across the sport.

The Lotus 49's story is one of innovation, success, and lasting impact. It is a testament to Colin Chapman's genius and his philosophy of "simplify, then add lightness." The car's legacy continues to be celebrated in the world of motorsports, remembered as a groundbreaking machine that reshaped the landscape of Formula 1 engineering.

Lotus 72 (1970-1975)

The Lotus 72, one of the most successful and iconic Formula 1 cars, competed from 1970 to 1975. Designed by Colin Chapman and Maurice Philippe for Team Lotus, the Lotus 72 was a revolutionary design that significantly influenced the future of Formula 1 car engineering.

Background and Development

- **Innovative Design**: The Lotus 72 was developed as a successor to the successful Lotus 49. It featured several innovative design elements, including side-mounted radiators, which reduced the car's frontal area and improved aerodynamic efficiency.

- **Focus on Aerodynamics**: The car was one of the first to take into account aerodynamic principles seriously. Its wedge-shaped design and inboard brakes were part of this aerodynamic focus.

Technical Features

- **Advanced Chassis**: The 72's chassis was a partially-stressed monocoque with an aluminum skin, which contributed to its light weight and structural rigidity.

- **Suspension Innovations**: The car had inboard front brakes and a torsion bar suspension, which helped lower the center of gravity and reduce unsprung weight, improving handling and grip.

Racing Success

- **Dominant Performance**: The Lotus 72 was highly successful on the track. It won the Constructors' Championship in 1970, 1972, and 1973.

- **Driver Championships**: It brought World Championships for Jochen Rindt, who won posthumously in 1970, and Emerson Fittipaldi in 1972.
- **Longevity and Adaptability**: The 72 remained competitive for an unusually long period in the fast-evolving world of Formula 1, with continuous development keeping it at the front of the grid.

Drivers

- **Jochen Rindt**: Rindt's performances in the 72 were exceptional, and his tragic death at Monza in 1970 came when he was leading the championship.
- **Emerson Fittipaldi**: Fittipaldi's driving in the 72 helped solidify his status as one of the sport's greats, winning his first championship at the age of 25.

Legacy and Impact

- **Innovative Impact**: The Lotus 72's innovative design influenced the engineering of future Formula 1 cars, especially in terms of aerodynamic design and radiator placement.

- **Iconic Status**: With its distinctive shape and racing success, the Lotus 72 became an iconic symbol of Formula 1 innovation and competitiveness in the early 1970s.

- **Enduring Influence**: The principles and design concepts introduced by the Lotus 72 continued to shape Formula 1 car design for many years, making it one of the most influential cars in the history of the sport.

The Lotus 72's story is a testament to the genius of Colin Chapman and his team's ability to continually evolve and adapt the car to maintain its competitive edge. It remains a celebrated example of innovation, performance, and success in Formula 1 history.

Ferrari 312T (1975-1980)

The Ferrari 312T series, which competed in Formula 1 from 1975 to 1980, is a hallmark in the storied history of Ferrari's racing legacy. Designed by Mauro Forghieri, the 312T was a series of cars that combined innovative design with the classic Ferrari power to dominate the sport during its tenure.

Background and Development

- **Revolutionizing Ferrari's Approach**: The 312T was developed to improve upon its predecessor, the 312B3. Ferrari aimed to resolve issues with handling and weight distribution that had hampered previous models.

- **Technical Innovation**: The "T" in 312T stood for "transversale," referring to the transverse mounting of the gearbox, which improved the car's balance and handling.

Design and Technical Features

- **Chassis and Aerodynamics**: The 312T series featured a tubular aluminum monocoque chassis. The design focused on achieving a low center of gravity and improved aerodynamics.

- **Powerful Flat-12 Engine**: The car was powered by a powerful and reliable flat-12 engine, which provided a significant advantage in terms of power and torque.

Racing Success

- **Dominant Performance**: The Ferrari 312T series was extraordinarily successful on the track. It won the Constructors' Championship in 1975, 1976, 1977, 1979, and 1980.

- **Driver Championships**: Niki Lauda won the World Championship in 1975 and 1977 driving the 312T. Jody Scheckter also claimed the championship in 1979 with the 312T4 model.

- **Consistent Evolution**: The 312T series saw various iterations, including the 312T2, T3, T4, and T5, each designed to adapt to changing regulations and competition.

Notable Drivers

- **Niki Lauda**: Lauda's skill and feedback were instrumental in developing the 312T. His championships in 1975 and 1977 were testaments to his driving prowess and the car's design.

- **Jody Scheckter**: Scheckter's 1979 championship with the 312T4 was the last Drivers' Championship for Ferrari until Michael Schumacher's win in 2000.

Legacy and Impact

- **Lasting Influence**: The 312T series' success helped revive Ferrari's fortunes in Formula 1, establishing a legacy that would continue in the decades to come.

- **Engineering Benchmark**: The series set a benchmark in terms of integrating engine and chassis design, influencing future F1 car development.

- **Iconic Status**: With its classic red livery and distinctive engine sound, the 312T series became an icon of Ferrari's racing heritage, embodying the passion and technical prowess of the Scuderia.

The story of the Ferrari 312T series is one of remarkable success, technical innovation, and racing dominance. It stands as a shining example of Ferrari's commitment to excellence in Formula 1 and remains a beloved chapter in the annals of motorsport history.

McLaren MP4/4 (1988)

The McLaren MP4/4, which competed in the 1988 Formula 1 season, is widely considered one of the most dominant and successful Formula 1 cars in the history of the sport. Designed by Steve Nichols and Gordon Murray for McLaren, the MP4/4 set new standards in efficiency, design, and performance.

Background and Development

- Pursuit of Perfection: The McLaren MP4/4 was developed to maintain McLaren's competitive edge in Formula 1. The team sought to create a car that was not only fast but also reliable and efficient.

- Innovative Design: The car's design was a significant departure from its predecessor, the MP4/3. It featured a lower, sleeker profile and a more aerodynamic shape, inspired partly by Murray's work on the Brabham BT55.

Design and Technical Features

- Lowline Aerodynamics: One of the key features of the MP4/4 was its "lowline" aerodynamic philosophy, which offered a lower center of gravity and reduced aerodynamic drag.

- **Honda Turbo Engine:** The car was powered by a highly efficient and powerful Honda RA168-E turbocharged V6 engine, known for its reliability and power output.

Racing Success

- **Unprecedented Dominance:** The McLaren MP4/4's performance during the 1988 season was unparalleled. The car won 15 out of the 16 races, a record for the highest winning percentage (93.75%) in a single season.
- **Ayrton Senna's First Championship:** Ayrton Senna, in his first year with McLaren, won his first World Championship in the MP4/4. He won 8 races, including memorable victories in Monaco and Japan.
- **Alain Prost's Performance:** Alain Prost also had a remarkable season, winning 7 races. Although he scored more points than Senna, the best 11 results rule meant Senna took the championship.

Notable Drivers

- **Ayrton Senna:** Senna's exceptional skill, particularly in qualifying, was a perfect match for the MP4/4's capabilities.
- **Alain Prost:** Prost's smooth and calculated driving style complemented the car's design, making the Senna-Prost pairing one of the most formidable in F1 history.

Legacy and Impact

- Benchmark in F1 History: The McLaren MP4/4 set a benchmark in Formula 1 for car design and team performance. Its dominance influenced the design philosophies of future F1 cars.

- Iconic Status: The car is remembered as an icon, not only for its performance but also for its striking red and white livery, representing the peak of the turbo era in Formula 1.

- End of an Era: The 1988 season marked the end of the turbocharged era in Formula 1, with naturally aspirated engines becoming mandatory from 1989 onwards.

The McLaren MP4/4's story is one of technological excellence, near-perfect execution, and historic achievement. It remains a pinnacle in the annals of Formula 1, epitomizing the combination of engineering ingenuity and driving mastery.

Williams FW14B (1992)

The Williams FW14B, which competed in the 1992 Formula 1 season, is widely regarded as one of the most technologically advanced and dominant Formula 1 cars ever built. Engineered by Patrick Head and Adrian Newey for the Williams team, the FW14B was a masterpiece of innovation and performance.

Background and Development

- **Evolution from FW14**: The FW14B was an evolution of the Williams FW14 from 1991. The team focused on enhancing its technological capabilities, particularly in the areas of active suspension, traction control, and semi-automatic transmission.

- **Technological Leap**: The car was at the forefront of applying sophisticated electronic technologies in Formula 1, which were groundbreaking at the time.

Design and Technical Features

- **Active Suspension**: One of the FW14B's key features was its active suspension system, which electronically controlled the ride height and damping in response to track conditions, improving grip and stability.

- **Advanced Aerodynamics**: The car also boasted advanced aerodynamics, which, when combined with the active suspension, resulted in exceptional handling and speed, particularly through corners.

- **Powerful Engine**: The FW14B was powered by a Renault V10 engine, which was among the most powerful and reliable in the field.

Racing Success

- **Dominant Season**: The 1992 season was dominated by the Williams FW14B. The car won 10 out of the 16 races, showcasing its superiority over the competition.

- **Nigel Mansell's Championship**: The car propelled Nigel Mansell to his only World Championship. Mansell set a then-record of 9 wins in a single season and clinched the title with five races to spare.

- **Team Success**: Williams also easily secured the Constructors' Championship, underlining the car's and the team's dominance.

Notable Drivers

- **Nigel Mansell**: Mansell's aggressive driving style was perfectly suited to the FW14B's capabilities. His record-breaking season included a then-record of 14 pole positions.

- **Riccardo Patrese**: The experienced Italian driver, Riccardo Patrese, also enjoyed success in the FW14B, finishing second in the championship.

Legacy and Impact

- **Pinnacle of Technology**: The FW14B represented the pinnacle of technological innovation in Formula 1 at the time. Its use of active suspension and other electronic systems set a new standard in the sport.

- **Regulation Changes**: The car's dominance and technological sophistication led to regulatory changes, as Formula 1 authorities sought to dial back the level of technological intervention in the sport.

- **Iconic Status**: The Williams FW14B remains an iconic Formula 1 car, remembered for its advanced technology, striking blue and white livery, and its overwhelming dominance during the 1992 season.

The story of the Williams FW14B is one of engineering excellence, technological innovation, and on-track dominance. It not only marked a high point in Williams' racing history but also left a lasting impact on the development and regulation of Formula 1 cars.

Benetton B194 (1994)

The Benetton B194, which competed in the 1994 Formula 1 season, is notable for its role in Michael Schumacher's first World Championship victory. Designed by Rory Byrne and Ross Brawn for the Benetton Formula team, the B194 was a car that combined innovative design with a controversial edge.

Background and Development

- **Targeting Championship Success**: The Benetton B194 was developed with the aim of making Benetton a title contender. The team focused on creating a car that was both fast and reliable, capitalizing on the regulation changes that banned active suspension and traction control systems.

- **Integration of Ford Engine**: The B194 was powered by a Ford Zetec-R V8 engine, which, while less powerful than some of its rivals' engines, was known for its light weight and good fuel efficiency.

Design and Technical Features

- **Aerodynamic Efficiency**: The B194 featured a distinctive aerodynamic design, with a narrow front profile to reduce drag. Its overall design emphasized efficient downforce generation.
- **Simplicity and Reliability**: In contrast to the complex active suspension systems of the previous year, the B194's more traditional suspension setup aimed to provide consistent performance and reliability.

Racing Success

- **Dominant Performance by Schumacher**: The 1994 season saw Michael Schumacher and the B194 dominate the early races. Schumacher won six of the first seven races, establishing a significant lead in the championship.
- **Controversies**: The B194 was involved in several controversies throughout the season, including allegations of illegal traction control and fuel irregularities. However, no conclusive evidence was ever found to penalize the team.

Notable Drivers

- **Michael Schumacher**: Schumacher's driving was a crucial factor in the success of the B194. His aggressive and skillful driving maximized the car's potential.
- **Jos Verstappen and JJ Lehto**: Schumacher's teammates, Verstappen and Lehto, also drove the B194 but with less success.

Legacy and Impact

- **Schumacher's First Championship**: The B194 is best remembered for being the car in which Michael Schumacher won his first World Championship, marking the beginning of his era of dominance in Formula 1.
- **Controversial Legacy**: The controversies surrounding the B194, particularly regarding the alleged use of illegal aids, have left a complex legacy. It remains a subject of debate and speculation in the Formula 1 community.
- **Influence on Regulations**: The B194's role in the 1994 season led to tighter scrutiny and changes in regulations to ensure fair competition and safety in the sport.

The Benetton B194's story is intertwined with both technical innovation and controversy. It stands as a significant chapter in Formula 1 history, highlighting the fine line between competitive advantage and the strict adherence to the sport's regulatory framework.

BENETTON B194

Ferrari F2004 (2004)

The Ferrari F2004, which competed in the 2004 Formula 1 season, is renowned as one of the most successful and dominant cars in the history of the sport. Designed by Rory Byrne and Ross Brawn for the Scuderia Ferrari team, the F2004 continued Ferrari's early-2000s dominance with remarkable efficiency and performance.

Background and Development

- **Culmination of Continuous Improvement**: The F2004 was the culmination of the development process that started with the earlier Ferrari F1 cars of the 2000s. It was designed to refine and improve upon the already successful F2003-GA.

- **Focus on Aerodynamics and Engine Performance**: The car featured numerous aerodynamic enhancements and a powerful engine, making it incredibly fast and reliable.

Design and Technical Features

- **Aerodynamic Efficiency**: The F2004 boasted highly refined aerodynamics, with a focus on maximizing downforce while minimizing drag, a key to its high-speed performance.
- **V10 Engine**: Powered by a 3.0-liter V10 engine, the F2004's power unit was known for its high power output and reliability, crucial in a season with strict engine regulations.

Racing Success

- **Dominant Season**: The 2004 season was one of the most dominant in Formula 1 history. The F2004 won 15 out of the 18 races, with Michael Schumacher winning 13 of those, a record for the most wins in a single season at the time.
- **Championship Titles**: Schumacher clinched his seventh and final World Championship in the F2004, while Ferrari easily secured the Constructors' Championship.

Notable Drivers

- **Michael Schumacher**: Schumacher's skill and experience were perfectly complemented by the F2004, making him virtually unbeatable throughout the season.
- **Rubens Barrichello**: The Brazilian driver also had a successful season, securing multiple wins and playing a key role in Ferrari's Constructors' title.

Legacy and Impact

- **Benchmark in F1 History**: The F2004 is often cited as one of the greatest Formula 1 cars ever built, setting a high benchmark in terms of speed, reliability, and success.
- **End of an Era**: The car's success marked the peak of Ferrari's and Schumacher's dominance in Formula 1, as regulatory changes in subsequent seasons would change the landscape of the sport.
- **Iconic Status**: With its distinctive red livery and impressive track record, the F2004 has earned an iconic status among Formula 1 cars, embodying the pinnacle of Ferrari's racing prowess in the early 21st century.

The Ferrari F2004's story is one of technological excellence, unmatched performance, and historical significance in the realm of Formula 1. Its legacy continues to be celebrated as a symbol of Ferrari's engineering mastery and Formula 1 dominance.

Red Bull RB9 (2013)

The Red Bull RB9, which competed in the 2013 Formula 1 season, is remembered as one of the most successful cars in the sport's recent history. Designed by Adrian Newey for Red Bull Racing, the RB9 was the last car to dominate the Formula 1 World Championship before the major regulation changes in 2014.

Background and Development

- **Continuation of Red Bull Dominance**: The RB9 was developed as a successor to the highly successful RB8. It was the final evolution of a design philosophy that had started with the RB5 in 2009.

- **Refinement Over Revolution**: The RB9 was more of a refinement than a revolution, building on the strengths of its predecessors while incorporating incremental improvements.

Design and Technical Features

- **Aerodynamic Efficiency**: The RB9, like its predecessors, featured extremely sophisticated aerodynamics. Adrian Newey's design focused on maximizing downforce while maintaining efficient airflow.

- **Renault V8 Engine**: The car was powered by a Renault RS27-2013 2.4-litre V8 engine. Although not the most powerful in the field, it was highly reliable and worked seamlessly with the car's aerodynamics.

Racing Success

- **Dominant Performance**: The 2013 season saw the RB9 dominate the grid. The car won 13 out of 19 races, with 12 of those victories coming from Sebastian Vettel.

- **Vettel's Fourth Championship**: The car propelled Sebastian Vettel to his fourth consecutive World Championship, a feat that solidified his status as one of the sport's greats.

- **Team Success**: Red Bull Racing also secured the Constructors' Championship with ease, marking their fourth consecutive title.

Notable Drivers

- **Sebastian Vettel**: Vettel's driving style perfectly complemented the RB9's characteristics. His ability to extract the maximum from the car, especially in qualifying, was key to his success.

- **Mark Webber**: The Australian driver, Mark Webber, also drove the RB9. While he didn't enjoy the same level of success as Vettel, he contributed significantly to the team's Constructors' Championship win.

Legacy and Impact

- **End of an Era**: The RB9 represented the peak and culmination of Red Bull's dominance in the V8 era of Formula 1. Its success marked the end of this particular technological era, as the sport moved to V6 turbo-hybrid engines in 2014.

- **Benchmark in Design**: The car stood as a benchmark in Formula 1 design, particularly in terms of aerodynamics. Newey's work on the RB9 was widely praised and studied within the Formula 1 community.

- **Influence on Regulations**: The dominance of the RB9 and its predecessors influenced the direction of future Formula 1 regulations, particularly those aimed at reducing the importance of aerodynamics in car performance.

The Red Bull RB9's story is one of technical excellence, on-track dominance, and the end of a significant era in Formula 1. It remains a prominent example of how innovative design and engineering can lead to unmatched success in the sport.

Mercedes F1 W07 Hybrid (2016)

The Mercedes F1 W07 Hybrid, which competed in the 2016 Formula 1 season, is renowned for its technological sophistication and dominance in the hybrid era of Formula 1. Developed by Mercedes AMG Petronas Formula One Team, the W07 Hybrid continued Mercedes' overwhelming success in the turbo-hybrid era that began in 2014.

Background and Development

- **Continued Evolution:** The W07 Hybrid was an evolution of its highly successful predecessors, the W05 and W06. The focus was on refining and improving the already successful design, particularly in terms of aerodynamics, chassis balance, and power unit efficiency.

- **Hybrid Power Unit:** A key component of the W07's success was its power unit, the Mercedes PU106C Hybrid. It was not only powerful but also highly reliable and efficient, a critical advantage in the era of hybrid technology in F1.

Design and Technical Features

- **Advanced Aerodynamics**: The W07 featured refined aerodynamics with an emphasis on reducing drag while maximizing downforce, crucial for its superior performance on a variety of circuits.

- **Integration of Power Unit and Chassis**: The car's design effectively integrated the hybrid power unit with the chassis, ensuring optimal balance and weight distribution.

Racing Success

- **Dominant Performance**: The W07 Hybrid dominated the 2016 season, winning 19 out of 21 races, a record at the time for the most wins in a single season.

- **Drivers' and Constructors' Championships**: The car secured both the Drivers' and Constructors' Championships for Mercedes. Nico Rosberg won the Drivers' Championship, narrowly beating teammate Lewis Hamilton.

- **Reliability and Consistency**: The W07's reliability was a key factor in its success, allowing Mercedes' drivers to consistently finish races in top positions.

Notable Drivers

- **Nico Rosberg**: Rosberg's consistency and performance in the W07 were key to his first and only World Championship. His rivalry with teammate Hamilton was one of the defining narratives of the season.

- **Lewis Hamilton**: Hamilton, despite facing some reliability issues and setbacks during the season, demonstrated his exceptional skill, winning 10 races.

Legacy and Impact

- **Benchmark in Hybrid Era**: The W07 Hybrid set a high benchmark in the hybrid era of Formula 1, showcasing the potential of hybrid technology in motorsport.

- **Influence on Future Designs**: The car's design and performance influenced the development of subsequent Formula 1 cars, particularly in terms of integrating hybrid power units with chassis design.
- **End of an Era for Rosberg**: The 2016 season marked the end of Rosberg's Formula 1 career, as he announced his retirement shortly after winning the championship.

The Mercedes F1 W07 Hybrid's story is one of technical innovation, racing dominance, and a pivotal moment in the turbo-hybrid era of Formula 1. Its success further cemented Mercedes' status as a powerhouse in modern Formula 1 and pushed the boundaries of hybrid technology in racing.

Red Bull Racing RB16B (2021)

The Red Bull Racing RB16B, which competed in the 2021 Formula 1 season, is notable for breaking Mercedes' seven-year streak of dominance and propelling Max Verstappen to his first World Championship. This car was an evolution of its predecessor, the RB16, and was developed by Red Bull Racing with a focus on addressing the shortcomings of the previous model and maximizing the potential of the new aerodynamic regulations.

Background and Development

- **Refinement of RB16**: The RB16B was a refined version of the 2020 car, the RB16. Red Bull focused on improving the car's balance and aerodynamic stability, which had been issues in the previous year.

- **Regulation Changes**: The car was designed in the context of new aerodynamic regulations for 2021, which included changes to the floor, brake ducts, and bargeboards to reduce downforce.

Design and Technical Features

- **Aerodynamic Improvements**: Despite the regulatory restrictions, the RB16B featured enhanced aerodynamics, with a focus on optimizing airflow and downforce, particularly at the rear of the car.

- **Honda Power Unit**: The RB16B was powered by a Honda engine, which proved to be both powerful and reliable, a crucial factor in the tight championship battle.

Racing Success

- **Challenging Mercedes**: The RB16B proved to be a formidable challenger to Mercedes, winning races and consistently finishing on the podium.

- **Verstappen's First Championship**: Max Verstappen, with his aggressive driving style and remarkable skill, capitalized on the car's strengths, winning his first World Championship in a dramatic and controversial final race at the Abu Dhabi Grand Prix.

- **Constructors' Championship Battle**: While Red Bull ultimately finished second in the Constructors' Championship, the RB16B brought them significantly closer to Mercedes compared to previous years.

Notable Drivers

- **Max Verstappen**: Verstappen's talent was a perfect match for the RB16B's capabilities, and his performances were key to the car's success.
- **Sergio Perez**: The experienced Mexican driver, Sergio Perez, joined Red Bull in 2021 and contributed valuable points to the team's Constructors' Championship campaign.

Legacy and Impact

- **Breaking Mercedes' Dominance**: The RB16B is remembered for ending Mercedes' consecutive run of World Championships, marking a shift in the competitive landscape of Formula 1.
- **Highlighting Honda's Contribution**: The car showcased the progress made by Honda as an engine supplier, contributing to their successful return to Formula 1.
- **Adaptation to Regulation Changes**: The RB16B demonstrated Red Bull Racing's ability to adapt effectively to significant aerodynamic regulation changes, underscoring the team's engineering prowess.

The Red Bull Racing RB16B's story is one of overcoming challenges, technical refinement, and competitive edge. It played a pivotal role in one of the most thrilling and closely contested seasons in recent Formula 1 history, marking a significant chapter in the sport's ongoing narrative.

Red Bull Racing RB19

The Red Bull Racing RB19, competing in the 2023 Formula One World Championship, has marked a significant chapter in Formula 1 history due to its dominant performance. Designed and constructed by Red Bull Racing, the car was unveiled in New York City on February 3, 2023, and it signifies the return of Honda as a named engine supplier to Red Bull Racing and AlphaTauri, with the engines badged as Honda RBPT.

Technical Specifications

- **Engine**: The RB19 is equipped with a Honda RBPTH001 engine, a 1.6-liter direct injection V6 turbocharged engine limited to 15,000 rpm. It also features a Honda electric motor, kinetic and thermal energy recovery systems, and a Honda Lithium-ion battery. The power output is an impressive 1,080 hp (805 kW).

- **Suspension**: The front suspension of the RB19 comprises multi-link pull-rod actuated dampers and an anti-roll bar.

Racing Performance

- **Dominant Season**: The RB19 has been one of the most dominant cars in Formula One history, winning 20 out of 21 races in the season it competed.

- **Championships**: This remarkable performance led to Red Bull Racing winning the Constructors' Championship and Max Verstappen, the defending world champion, securing the Drivers' Championship in 2023.

The RB19, driven by Max Verstappen and Sergio Pérez, has showcased a combination of technical innovation and racing prowess, reinforcing Red Bull Racing's status as a top competitor in the Formula One World Championship.

Chapter 3:

The Constructors

Scuderia Ferrari

Scuderia Ferrari, the most storied and successful team in Formula 1 history, has been an integral part of the sport since its inception. Founded by Enzo Ferrari in 1929, the team initially focused on racing Alfa Romeos before building its first car in 1947.

Early Years in Formula 1

- **Entry into Formula 1**: Ferrari entered Formula 1 in the first World Championship in 1950 and won its first race in 1951 at the British Grand Prix.

- **First Championship Win**: The team won its first drivers' championship with Alberto Ascari in 1952 and 1953.

Era of Expansion and Success

- **1960s and 1970s**: Ferrari continued to grow in the 1960s and 1970s, winning multiple championships with drivers like John Surtees, Niki Lauda, and Jody Scheckter.
- **Innovations**: Ferrari was known for its innovative approach, pioneering developments in car design and aerodynamics.

Challenges and Comebacks

- **Periods of Struggle**: Despite their early success, Ferrari faced periods of struggle, particularly in the late 1980s and early 1990s.
- **Schumacher Era**: The signing of Michael Schumacher in 1996 marked the beginning of a dominant era. Under the leadership of Jean Todt, with Ross Brawn and Rory Byrne, Ferrari won five consecutive drivers' titles from 2000 to 2004.

Recent Years

- **Continued Competitiveness**: In the 2010s, Ferrari remained competitive, often challenging for championships with drivers like Fernando Alonso and Sebastian Vettel.
- **Current Era**: In recent years, Ferrari has been working towards returning to the top, focusing on developing competitive cars and nurturing talent like Charles Leclerc and Carlos Sainz Jr.

Legacy

- **Iconic Status**: Ferrari is not just a racing team but a symbol of Italian passion and excellence in motorsport. Their prancing horse logo is iconic, symbolizing speed, strength, and elegance.
- **Contribution to F1**: With the most wins, podium finishes, and championships in the history of the sport, Scuderia Ferrari's contribution to Formula 1 is unparalleled.

Scuderia Ferrari's story is intertwined with the history of Formula 1 itself, representing a legacy of passion, innovation, and excellence in motorsport.

McLaren Racing

McLaren Racing, founded in 1963 by New Zealander Bruce McLaren, is one of the most successful teams in Formula 1 history.

Early Success

- **Founding and Early Years**: The team entered Formula 1 in 1966. Bruce McLaren himself scored the team's first victory at the 1968 Belgian Grand Prix.

- **1970s Growth**: After Bruce McLaren's tragic death in 1970, the team continued to grow, winning races and championships.

The Ron Dennis Era

- **Ron Dennis' Takeover**: In 1981, Ron Dennis took over the team and transformed it into a highly professional organization.

- **Partnership with TAG and Honda**: The 1980s and 1990s saw McLaren reach new heights with TAG-Porsche and Honda engines, winning multiple championships with drivers like Niki Lauda, Alain Prost, and Ayrton Senna.

Later Success and Challenges

- **Mercedes Partnership**: A partnership with Mercedes-Benz in the late 1990s and 2000s led to more success, including championships with Mika Häkkinen and Lewis Hamilton.

- **Recent Years**: The 2010s were more challenging for McLaren, with a notable struggle during their partnership with Honda. However, recent seasons have seen a resurgence in performance.

Innovation and Impact

- **Technological Innovations**: McLaren has been known for its technological innovations, including pioneering carbon-fiber composite chassis construction.

- **Legacy**: Beyond F1, McLaren has expanded into road car production and other racing ventures, becoming a globally recognized brand.

McLaren's story in F1 is one of innovation, success, and resilience, marked by periods of dominance and challenges, but always pushing the boundaries of technology and performance.

Williams Racing

Williams Racing, officially founded as Williams Grand Prix Engineering in 1977 by Sir Frank Williams and Patrick Head, is one of the most respected and successful teams in Formula 1 history.

Rise to Prominence

- **Early Success**: Williams quickly rose to prominence, winning their first race in 1979 and their first drivers' and constructors' championships in 1980.

- **1980s and 1990s Dominance**: The team enjoyed great success in the 1980s and 1990s, winning multiple championships with drivers like Alan Jones, Keke Rosberg, Nelson Piquet, Nigel Mansell, Alain Prost, Damon Hill, and Jacques Villeneuve.

Technological Innovations

- **Pioneering Technology**: Williams was known for its engineering prowess, pioneering active suspension and other innovations.

Challenges and Resilience

- **Financial and Competitive Challenges**: In the 2000s and 2010s, Williams faced financial and competitive challenges but remained committed to racing in Formula 1.

- **Recent Developments**: The team has undergone significant changes, including new ownership, as it works to return to the front of the grid.

Williams Racing's story is one of passion, innovation, and excellence in Formula 1, marked by remarkable triumphs and resilience in the face of challenges.

Mercedes-AMG Petronas Formula One Team

The Mercedes-AMG Petronas Formula One Team, commonly known as Mercedes, is one of the most successful teams in the history of Formula One racing.

Origins:

- The team traces its roots back to the 1950s when it was known as Daimler-Benz AG and later became Mercedes-Benz in Formula One.

- The modern Mercedes works team emerged in 2010 when Mercedes-Benz acquired the Brawn GP team, marking their return to Formula One racing.

Dominance with the Hybrid Era:

- Mercedes became the dominant force in Formula One during the hybrid era, which began in 2014 with the introduction of turbocharged V6 hybrid power units.

- Led by team principal Toto Wolff and technical director James Allison, Mercedes consistently showcased innovation and performance.

Championships and Records:

- Mercedes has achieved unprecedented success, winning multiple Constructors' and Drivers' Championships.

- Lewis Hamilton, one of their star drivers, has set numerous records, becoming one of the most successful drivers in Formula One history.

Key Personnel:

- The team has had talented drivers like Lewis Hamilton, Nico Rosberg, and Valtteri Bottas.
- Figures like Ross Brawn, Paddy Lowe, and James Allison have contributed to the development of championship-winning cars.

Sustainability and Innovation:

- Mercedes has been at the forefront of sustainability and innovation in Formula One, actively pushing for environmentally friendly technologies.
- Their partnership with PETRONAS focuses on developing advanced fuels and lubricants for improved performance and efficiency.

Challenges and Rivalries:

- Mercedes has faced strong competition and rivalries with teams like Scuderia Ferrari and Red Bull Racing, adding drama and excitement to Formula One seasons.

The Mercedes-AMG Petronas Formula One Team's story is one of modern dominance, innovation, and success in the world of Formula One racing. They have established themselves as a powerhouse in the sport, shaping the future of Formula One.

Red Bull Racing

Red Bull Racing, a prominent name in Formula One, emerged on the racing scene with a burst of energy in the early 2000s. Founded through the acquisition of the Jaguar Racing team by Red Bull GmbH, the team quickly made a name for itself with its dynamic and aggressive approach to racing. Over the years, they have enjoyed periods of dominance, faced fierce competition, and seen star drivers like Sebastian Vettel and Max Verstappen don their colors. This is the story of Red Bull Racing's journey through the high-speed world of Formula One.

Origins:

- Red Bull Racing, also known as Aston Martin Red Bull Racing due to a sponsorship deal, is a Formula One racing team.

- The team originated from the purchase of the Jaguar Racing team by Red Bull GmbH in 2004.

Rise to Prominence:

- Red Bull Racing quickly rose to prominence with the infusion of Red Bull's energy drink marketing and financial support.
- The team's dynamic and aggressive approach to racing attracted attention.

Dominance with Sebastian Vettel:

- Red Bull Racing enjoyed a period of dominance with driver Sebastian Vettel in the early 2010s.

- Vettel won four consecutive Drivers' Championships from 2010 to 2013, and the team secured multiple Constructors' Championships.

Transition to Hybrid Era:

- With the introduction of the hybrid power unit era in 2014, Red Bull Racing faced challenges from Mercedes and Ferrari.
- They continued to be competitive but faced tougher competition.

Max Verstappen Era:

- Dutch driver Max Verstappen joined Red Bull Racing and became the team's star driver.
- Verstappen's aggressive driving style and talent brought renewed excitement to the team.

Key Personnel:

- Key figures in Red Bull Racing include team principal Christian Horner and technical advisor Adrian Newey.
- Their collaborative efforts have contributed to the team's success.

Sponsorship and Branding:

- Red Bull Racing's partnership with Red Bull energy drink has been central to their branding and marketing strategy.
- The team's distinctive livery and association with Red Bull have made them easily recognizable.

Challenges and Rivalries:

- Red Bull Racing has had intense rivalries with teams like Mercedes and Ferrari, leading to thrilling competitions on the track.

- Their battles for championships have been a highlight of recent Formula One seasons.

Red Bull Racing's story is marked by its rapid ascent in Formula One, periods of dominance, and a strong focus on branding and marketing. With talented drivers and key personnel, the team continues to be a competitive force in the sport.

Lotus F1 Team

Lotus F1 Team, a name steeped in Formula One history, has undergone various incarnations and transformations over the years. Founded as Toleman Motorsport in the 1980s, the team later became Benetton Formula and Renault F1 Team before adopting the Lotus name. With a rich heritage and a legacy of iconic cars and legendary drivers, Lotus F1 Team has left an indelible mark on the sport. This is the story of their journey through the fast-paced world of Formula One.

Origins:

- The team's roots can be traced back to Toleman Motorsport, founded in the early 1980s.

- Over the years, the team underwent several changes in ownership and names, including Benetton Formula and Renault F1 Team.

Lotus Revival:

- In 2012, the team was renamed Lotus F1 Team after a sponsorship agreement with Lotus Cars, reviving the historic Lotus name in Formula One.

- The team aimed to capture the spirit of the legendary Lotus cars that had competed in Formula One in the past.

Notable Drivers:

- Lotus F1 Team has been home to several iconic drivers, including Ayrton Senna, Michael Schumacher, and Fernando Alonso during their Benetton and Renault eras.

Challenges and Achievements:

- The team faced challenges in the highly competitive world of Formula One, but they also achieved notable successes, including multiple

Drivers' and Constructors' Championships during their Benetton and Renault periods.

Rebranding to Alpine:

- In 2021, the team underwent another transformation and was rebranded as Alpine F1 Team, focusing on performance and technology transfer from the Alpine road car division.

Heritage and Legacy:

- Lotus F1 Team's legacy is intertwined with the storied history of Lotus cars in Formula One, known for their innovative designs and racing pedigree.

Innovation and Engineering:

- The team has a history of innovation and engineering excellence, with technical developments that have had a lasting impact on the sport.

Lotus F1 Team, with its storied history and iconic name, has been a part of Formula One's evolution, showcasing innovation, legendary drivers, and a rich racing heritage. Though the team has seen different chapters, its legacy endures in the world of motorsport.

Renault F1 Team

Renault F1 Team, a prominent presence in Formula One, has a storied history that spans several decades. Founded in the late 1970s as a full-fledged works team by the French automobile manufacturer Renault, the team has gone through various phases of success and transformation. With multiple Constructors' and Drivers' Championships to their name, Renault F1 Team has been a force to be reckoned with in the world of motorsport. This is the story of their journey through the high-speed world of Formula One.

Origins and Early Success:

- Renault entered Formula One as a constructor in 1977 and quickly made their mark with turbocharged engines.

- They achieved early success with drivers like Alain Prost and René Arnoux, showcasing the potential of turbo power.

Constructors' Championships:

- Renault F1 Team secured their first Constructors' Championship in 1992, followed by another in 1993.
- Their partnership with Williams during this period was particularly fruitful.

Benetton and Alonso Era:

- The team was rebranded as Benetton Formula in the mid-1990s but continued to be competitive.

- The late 2000s saw the emergence of Renault F1 Team as a powerhouse, with Fernando Alonso winning back-to-back Drivers' Championships in 2005 and 2006.

Transition to Lotus:

- The team transitioned to the Lotus F1 Team in 2012, retaining a connection to the Renault brand as an engine supplier.
- During this period, the team showcased its engineering prowess and driver talent.

Return to Renault F1 Team:

- In 2016, the team reverted to the Renault F1 Team name as Renault made a full-fledged return as a works team.
- Their goal was to challenge the top teams and build on their past success.

Challenges and Revival:

- The team faced challenges but gradually worked towards improving competitiveness with a focus on developing their own power unit.

Alpine Rebranding:

- In 2021, the team underwent another transformation and was rebranded as Alpine F1 Team, aligning with the Alpine road car brand.
- This marked a new chapter for the team with an emphasis on performance and innovation.

Heritage and Legacy:

- Renault F1 Team's legacy is marked by its contributions to Formula One's technological advancements and its role in nurturing driver talent.

Current Endeavors:

- Alpine F1 Team continues to compete in Formula One, combining the expertise of Renault with the heritage of Alpine cars to aim for success in the sport.

Renault F1 Team's journey through Formula One has been marked by innovation, championships, and a commitment to racing excellence. Their ability to adapt and evolve over the years has solidified their place in the history of motorsport.

Alfa Romeo Racing

Alfa Romeo Racing, a name synonymous with motorsport heritage and Italian flair, has a rich history in Formula One. With origins dating back to the early days of Grand Prix racing, the team has undergone various transformations, partnerships, and rebranding efforts. Alfa Romeo's return to Formula One in the modern era has brought a touch of nostalgia and excitement to the sport. This is the story of Alfa Romeo Racing's journey through the fast-paced world of Formula One.

Early Grand Prix Racing:

- Alfa Romeo's involvement in motorsport dates back to the early 20th century, with notable success in Grand Prix racing.

- They achieved victories and championships in the pre-Formula One era, cementing their place in racing history.

Reentry into Formula One:

- Alfa Romeo made a return to Formula One in 1950 when they supplied engines and cars to various teams.

- Their cars, often featuring the iconic Quadrifoglio (four-leaf clover) emblem, became a symbol of Italian racing excellence.

1970s Success:

- The 1970s saw Alfa Romeo compete as a full-fledged factory team, achieving some success with drivers like Niki Lauda and Mario Andretti.

- They secured a few Grand Prix wins during this period.

Partnerships and Rebranding:

- Alfa Romeo's presence in Formula One has been associated with various teams, including March, Osella, and Sauber.

- The team has undergone rebranding efforts, such as Sauber Alfa Romeo, to align with Alfa Romeo's vision of motorsport.

Return as Alfa Romeo Racing:

- In 2019, the team returned as Alfa Romeo Racing, forming a partnership with Sauber Motorsport.
- The Alfa Romeo name reentered Formula One with a focus on nurturing young talent and competitive racing.

Drivers and Development:

- Alfa Romeo Racing has featured talented drivers like Kimi Räikkönen and Antonio Giovinazzi.
- The team has also played a role in the development of young drivers through their junior program.

Challenges and Ambitions:

- Alfa Romeo Racing has faced challenges in the highly competitive Formula One landscape but remains committed to achieving success.
- Their ambitions include securing podium finishes and building on their rich motorsport legacy.

Italian Heritage and Passion:

- Alfa Romeo Racing embodies the passion and racing spirit of Italian motorsport, with a strong connection to Alfa Romeo's road car heritage.

Current Endeavors:

- Alfa Romeo Racing continues to compete in Formula One, representing the Alfa Romeo brand on the global stage and keeping the tradition of Italian racing alive.

Alfa Romeo Racing's journey in Formula One is a tale of historical significance, reentry into modern motorsport, and a commitment to upholding the legacy of Italian racing excellence. Their presence on the grid adds a touch of nostalgia and passion to the world of Formula One.

Chapter 4:
The Iconic Tracks

Monaco Circuit (Circuit de Monaco)

The Monaco Circuit, officially known as the Circuit de Monaco, is one of the most iconic and historic circuits in Formula One racing. Located in the picturesque city of Monte Carlo, Monaco, the circuit's history is intertwined with the glamour and prestige of the principality. Here's a brief history behind the Monaco Circuit:

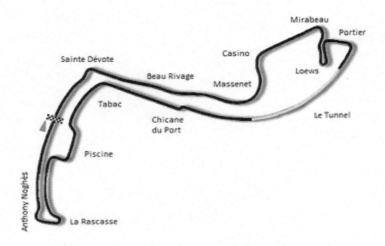

- ➢ **Inception and Early Years:** The Monaco Grand Prix, first held in 1929, predates the official Formula One World Championship. It was originally conceived as a way to promote the Monaco Grand Hotel and attract international attention to Monaco. The race was an instant success, featuring a challenging street circuit that wound its way through the narrow, winding streets of Monte Carlo.

- ➢ **Monaco in the Formula One World Championship:** When the Formula One World Championship was inaugurated in 1950, the Monaco Grand Prix was one of the inaugural races. Since then, it has been a regular fixture on the Formula One calendar, earning its reputation as one of the crown jewels of motorsport.

➢ **Challenging Layout:** The Monaco Circuit is known for its unique and demanding layout. It is a tight and twisty street circuit that features narrow roads, sharp corners, and elevation changes. The circuit is notorious for its lack of overtaking opportunities, making grid position and qualifying crucial for success.

➢ **Glamour and Prestige:** The Monaco Grand Prix is not just a race; it's a social event and a symbol of luxury and glamour. The race attracts celebrities, royalty, and high-profile personalities who flock to Monaco to witness the spectacle. The backdrop of the Mediterranean Sea, the Casino de Monte-Carlo, and the luxurious yachts in the harbor add to the event's allure.

➢ **Legendary Moments:** The Monaco Circuit has witnessed numerous legendary moments in Formula One history. Drivers like Ayrton Senna, who won the race a record six times, and Graham Hill, known as "Mr. Monaco" for his five wins, have left their mark on the circuit's history. Senna's duel with Alain Prost in 1984 and his extraordinary pole laps are some of the most iconic moments in Formula One history.

➤ **Modern Adaptations:** Over the years, the circuit has undergone minor modifications to improve safety and accommodate the evolving needs of Formula One racing. However, the essence of the track, with its tight and challenging nature, has remained largely unchanged.

➤ **Ongoing Prestige:** The Monaco Grand Prix continues to be a highlight of the Formula One season, attracting fans and teams from around the world. Its unique character, the allure of Monte Carlo, and the challenge it presents to drivers ensure that it remains a standout event on the calendar.

In summary, the Monaco Circuit's history is a tale of glamour, prestige, and legendary racing moments. It has been a cornerstone of Formula One racing since its early days and continues to captivate fans with its challenging and iconic street circuit.

Monza Circuit (Autodromo Nazionale Monza)

The Monza Circuit, officially known as the Autodromo Nazionale Monza, is one of the oldest and most iconic circuits in Formula One racing. Located near the city of Monza in Italy, this historic track has a rich and storied history that has made it a favorite among drivers and fans alike. Here's a brief history behind the Monza Circuit:

- ➢ **Early Beginnings:** The Monza Circuit was built in 1922 and hosted its first Italian Grand Prix in the same year. It was one of the original venues in the inaugural 1950 Formula One World Championship season, solidifying its place in motorsport history.

➤ **High-Speed Racing:** Monza is known for its high-speed straights and flowing curves. The circuit has traditionally featured a mix of high-speed sections and chicanes, making it a challenging track for both drivers and engineers.

➤ **Temple of Speed:** Monza is often referred to as the "Temple of Speed" due to its reputation as one of the fastest tracks on the Formula One calendar. The long straights, notably the iconic Parabolica and the old banked circuit, allowed cars to reach blistering speeds.

➤ **Historic Significance:** Monza has a special place in Formula One history as the venue for the first-ever Formula One World Championship race in 1950. The circuit has witnessed countless historic moments, including legendary battles and dramatic victories.

➤ **Monza Banking:** The circuit's old high-speed banking, known as the "Curva Sud," was once a defining feature of the track. While it is no longer used for Formula One races due to safety concerns, its legacy lives on in the history of Monza.

- ➤ **Tifosi and Ferrari:** Monza is Ferrari's home race, and the passionate Italian fans, known as the "Tifosi," flock to the circuit in red to support the Scuderia Ferrari team. The atmosphere during the Italian Grand Prix is electric, and the Tifosi's enthusiasm is a hallmark of the event.

- ➤ **Record-Breaking Moments:** Monza has seen numerous record-breaking moments in Formula One. Michael Schumacher's remarkable pole lap in 2003, with an average speed of over 162 mph (260 km/h), is one of the standout records.

- ➤ **Challenging Variations:** Over the years, Monza has undergone various modifications to enhance safety, but it has retained its essential character as a fast and flowing circuit. Variations of the layout have included the use of chicanes, such as the Variante Ascari and the Variante della Roggia.

- ➤ **Modern Era:** Monza continues to host the Italian Grand Prix and remains a fan favorite. It is a track where slipstreaming and daring overtaking maneuvers are common due to its long straights.

- ➤ **Heritage and Legacy:** The Monza Circuit's heritage and legacy in Formula One are undeniable. Its status as one of the classic tracks on the calendar ensures that it will always be an integral part of the sport's history.

In summary, the Monza Circuit is a historic and legendary venue in Formula One, known for its high-speed nature, passionate fans, and historic significance. It remains a beloved track that embodies the essence of speed and racing excellence in the world of motorsport.

Silverstone Circuit

The Silverstone Circuit, located in the United Kingdom, is one of the oldest and most iconic circuits in Formula One racing. Known for its high-speed layout and historic significance, Silverstone has played a pivotal role in the history of motorsport. Here's a brief history behind the Silverstone Circuit:

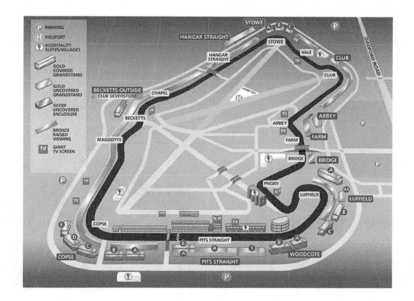

- **WWII Airfield Origins:** Silverstone's history as a racing circuit began during World War II when it was originally constructed as a Royal Air Force airfield. The airfield was used for bomber training and later became a base for the United States Army Air Forces.

- **Post-War Transformation:** After the war, Silverstone was repurposed as a racing circuit, making use of the existing runways and perimeter roads. The first official race took place in 1947, and the circuit quickly gained a reputation for its fast and challenging layout.

➢ **Inaugural World Championship Race:** In 1950, Silverstone hosted the first-ever Formula One World Championship race, marking the beginning of the modern era of Formula One. The event became known as the British Grand Prix and has been a fixture on the Formula One calendar since.

➢ **High-Speed Circuit:** Silverstone is renowned for its high-speed straights and sweeping corners, making it a favorite among drivers who appreciate its flowing nature. The fast and demanding track challenges both man and machine.

- **Historic Moments:** The circuit has witnessed numerous historic moments in Formula One, including the legendary rivalry between Alain Prost and Ayrton Senna, which produced memorable battles at Silverstone in the late 1980s.

- **Home of British Motorsport:** Silverstone is often referred to as the "Home of British Motorsport" and is the primary testing and development location for many Formula One teams based in the UK. The country's rich motorsport heritage is showcased at Silverstone.

- **Variations and Upgrades:** Over the years, Silverstone has undergone several changes and upgrades to improve safety and accommodate the evolving needs of Formula One racing. These modifications have maintained the circuit's character while enhancing its safety standards.

- **Fan-Friendly Atmosphere:** Silverstone is known for its enthusiastic and knowledgeable fans who create a vibrant and electric atmosphere during the British Grand Prix. The fans' passion and loyalty to their favorite drivers add to the event's charm.

- **Modern Era:** Silverstone continues to host the British Grand Prix and has also featured as part of the Formula One calendar for the 70th Anniversary Grand Prix. It remains a popular destination for both fans and teams.

- **Legacy and Heritage:** The Silverstone Circuit's legacy in Formula One is unparalleled. Its historical significance, fast-paced racing, and the iconic

Silverstone Wing complex ensure that it remains a cornerstone of motorsport history.

In summary, the Silverstone Circuit is a historic and iconic venue in Formula One, known for its high-speed challenges, historic moments, and passionate fans. It embodies the spirit of British motorsport and holds a special place in the hearts of racing enthusiasts around the world.

Spa-Francorchamps Circuit

The Spa-Francorchamps Circuit, located in the Ardennes forest of Belgium, is one of the most iconic and beloved circuits in Formula One history. Renowned for its challenging and scenic layout, Spa-Francorchamps has a rich heritage and has been the backdrop for countless memorable races. Here's a brief history behind the Spa-Francorchamps Circuit:

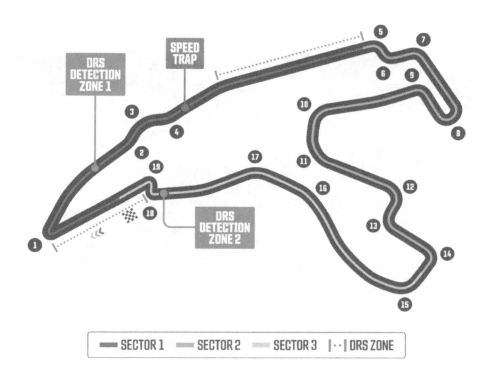

> **Historic Origins:** The circuit's origins date back to the early 1920s when the region of Spa-Francorchamps was already known for its natural springs and therapeutic baths. The idea to create a racing circuit in the area was conceived as a way to attract tourism.

- **High-Speed Layout:** Spa-Francorchamps is famous for its fast and flowing layout, characterized by long straights, sweeping curves, and dramatic elevation changes. The circuit's length and challenging corners make it one of the longest tracks on the Formula One calendar.

- **Eau Rouge-Raidillon:** One of the circuit's most iconic features is the Eau Rouge-Raidillon complex, a series of corners that climb steeply uphill. Drivers approach it at high speed and experience significant g-forces as they navigate this famous section.

- **Historical Significance:** Spa-Francorchamps hosted its first Formula One race in 1950, just a year after the inaugural World Championship season. Since then, it has become a cornerstone of Formula One racing and a favorite among drivers.

- **Unpredictable Weather:** The Ardennes region is known for its unpredictable weather, and rain can turn a race at Spa-Francorchamps into a chaotic and thrilling spectacle. The ever-changing conditions add an extra layer of excitement and challenge.

- **Legendary Moments:** The circuit has witnessed numerous legendary moments in Formula One history. Memorable battles, like the one between Ayrton Senna and Alain Prost in 1987, have etched their place in the annals of motorsport history.

- **Safety Improvements:** Over the years, Spa-Francorchamps has undergone safety improvements while preserving its classic character. Changes to the Bus Stop chicane and the addition of TecPro barriers are examples of efforts to enhance safety.

- **Fan-Friendly Atmosphere:** The Belgian Grand Prix at Spa-Francorchamps draws fans from around the world who appreciate the circuit's challenging nature and picturesque surroundings. The passionate and knowledgeable spectators create a unique atmosphere.

- **Modern Era:** Spa-Francorchamps remains a regular fixture on the Formula One calendar, hosting the Belgian Grand Prix. It is a track that continues to challenge drivers and captivate fans with its blend of speed and spectacle.

- **Heritage and Legacy:** The Spa-Francorchamps Circuit's heritage and legacy in Formula One are unrivaled. It embodies the spirit of classic Grand Prix racing and serves as a reminder of the sport's historic roots.

In summary, the Spa-Francorchamps Circuit is a historic and iconic venue in Formula One, known for its challenging layout, unpredictable weather, and legendary moments. It is a track that combines speed, skill, and natural beauty, making it a beloved destination for motorsport enthusiasts worldwide.

Suzuka Circuit

The Suzuka Circuit, located in Japan, is one of the most iconic and unique circuits in Formula One racing. Famous for its figure-eight layout and challenging corners, Suzuka has been the stage for many historic moments in the sport. Here's a brief history behind the Suzuka Circuit:

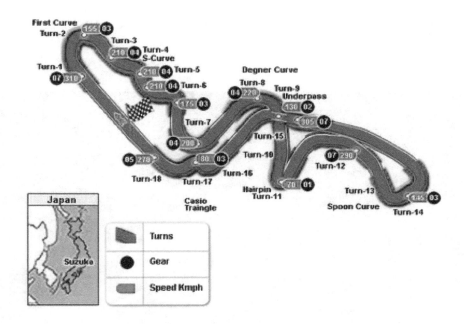

- ➤ **Designed by a Legend:** Suzuka Circuit was designed by Dutch architect and racetrack designer John Hugenholtz. His vision for the circuit was to create a challenging track that would test drivers' skills and provide exciting racing.

- ➤ **Unique Figure-Eight Layout:** Suzuka is one of the few circuits in the world with a figure-eight layout, meaning that the track crosses over itself. This layout adds to the circuit's character and complexity.

- **Inaugural Japanese Grand Prix:** Suzuka hosted its first Japanese Grand Prix in 1987. Since then, it has been a regular fixture on the Formula One calendar and has often been the venue for championship-deciding races.

- **Legendary S Curves:** The "S Curves" section at the beginning of the lap is one of the most iconic parts of the Suzuka Circuit. It challenges drivers with a series of fast and sweeping corners that require precision and control.

- **Senna-Prost Rivalry:** Suzuka witnessed one of the most intense rivalries in Formula One history between Ayrton Senna and Alain Prost. Their famous collisions and battles at the circuit added to its legendary status.

- **Championship Deciders:** Suzuka has been the venue for many championship-deciding races, creating dramatic and historic moments. The circuit's demanding nature often puts the title contenders to the ultimate test.

- **Technical and High-Speed:** The Suzuka Circuit combines technical sections with high-speed straights and corners. It challenges teams and drivers to find the right balance between downforce and top speed.

- **Japanese Passion:** Japanese fans are known for their passion and dedication to Formula One, and the Japanese Grand Prix at Suzuka attracts enthusiastic spectators from around the world.

- **Modern Era:** Suzuka remains a significant part of the Formula One calendar, hosting the Japanese Grand Prix. It is a track that continues to provide thrilling races and showcase the talents of drivers.

- **Legacy and Heritage:** The Suzuka Circuit's legacy in Formula One is marked by its unique layout, challenging corners, and historic moments. It is a circuit that stands out as a true driver's track.

In summary, the Suzuka Circuit is a unique and iconic venue in Formula One, known for its figure-eight layout, technical challenges, and its place in the sport's history. It is a circuit that has witnessed legendary battles and continues to captivate fans with its distinctive character and thrilling races.

Circuit of the Americas (COTA)

American Ambitions: COTA was designed with the goal of bringing Formula One back to the United States and establishing a world-class racing facility. The circuit was officially opened in 2012, making it one of the newest additions to the Formula One calendar.

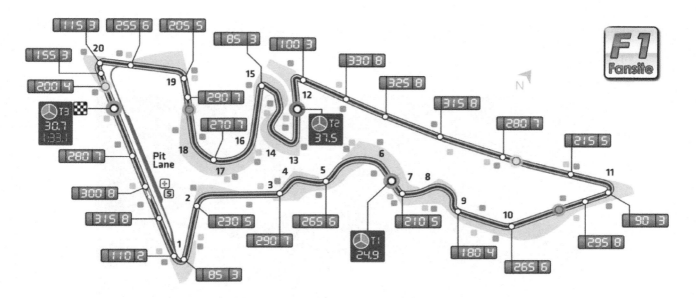

- **State-of-the-Art Facilities:** COTA boasts state-of-the-art facilities, including a purpose-built pit and paddock complex, a 251-foot observation tower, and a stunning amphitheater. These amenities make it a top-tier venue in the world of motorsport.

- **Diverse Layout:** The circuit features a diverse layout that incorporates elements inspired by iconic tracks from around the world. It includes challenging corners, elevation changes, and a long back straight that allows for exciting overtaking opportunities.

- **Host of the United States Grand Prix:** COTA has been the host of the United States Grand Prix since its inaugural race in 2012. The return of Formula One to the United States has been celebrated by fans and marked a significant moment in the sport's history.

- **Entertainment Hub:** Beyond Formula One, COTA has established itself as a year-round entertainment hub, hosting a variety of motorsport events, music festivals, and cultural activities. This multifaceted approach has contributed to its popularity.

- **Dramatic Races:** COTA has produced its fair share of memorable races, including exciting duels and strategic battles. The track's varied layout allows for different racing lines and tactics.

- **Signature Features:** The "Esses" section at the start of the lap and the challenging Turn 1 hairpin, inspired by the Hockenheimring, are signature features that make COTA unique and challenging for drivers.

- **Fan-Friendly Environment:** COTA is known for its fan-friendly atmosphere, with plenty of fan zones, entertainment, and activities throughout the race weekend. The circuit's location in Austin, often called the "Live Music Capital of the World," adds to the vibrant ambiance.

- **Modern Era:** COTA has quickly become a modern classic on the Formula One calendar, offering a fresh perspective on the sport in the United States.

- **Global Appeal:** COTA's international appeal has attracted fans from around the world, contributing to the globalization of Formula One and the growth of its fanbase.

Interlagos Circuit (Autódromo José Carlos Pace)

The Interlagos Circuit, officially known as the Autódromo José Carlos Pace, is a historic and challenging circuit located in São Paulo, Brazil. It has been a staple of the Formula One calendar for many years and has played host to numerous thrilling races. Here's a brief history behind the Interlagos Circuit:

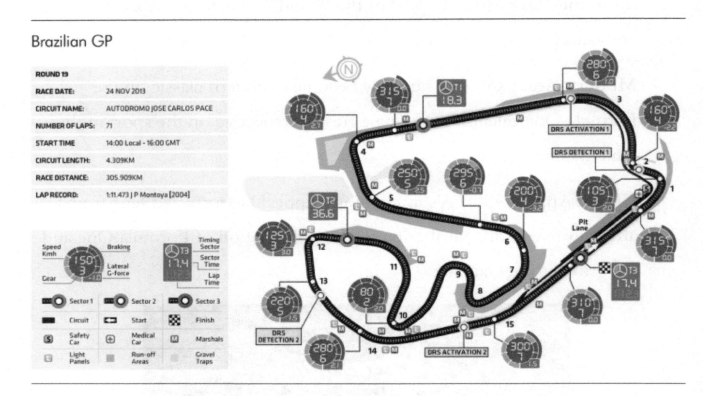

> ➤ **Origins and Renaming:** The circuit was originally built in the 1930s and initially known as the "Interlagos Circuit" due to its location between two large reservoirs, Guarapiranga and Billings. It was later renamed in honor of Brazilian Formula One driver José Carlos Pace, who tragically lost his life in a plane crash.

- **Early Brazilian Grand Prix:** The Brazilian Grand Prix has a long history dating back to the 1970s. Interlagos became the permanent home of the race in the late 1970s, making it an integral part of Formula One.

- **Challenging Layout:** Interlagos is known for its challenging and undulating layout. The circuit features a mix of high-speed straights and a technical infield section with tight corners, providing a well-rounded test for drivers and teams.

- **Rain and Unpredictability:** São Paulo's tropical climate often brings rain and unpredictable weather to the race, creating exciting and unpredictable races. The rain can turn Interlagos into a thrilling spectacle with changing conditions.

- **Historic Moments:** Interlagos has witnessed numerous historic moments in Formula One history. Memorable races, including championship-deciding showdowns, have unfolded at the circuit.

- **Senna's Home:** Interlagos is closely associated with the legendary Brazilian driver Ayrton Senna. His performances at the circuit, including his emotional maiden Formula One victory in 1991, have left an indelible mark on the track's history.

- **Spectacular Overtaking:** The circuit's layout and elevation changes provide ample opportunities for overtaking, making it a favorite among drivers and fans for exciting races.

- **Passionate Brazilian Fans:** Brazilian Formula One fans, known for their passion and support of their homegrown talent, add to the electric atmosphere at Interlagos. The "S" bleachers are famous for their enthusiastic fans.

- **Safety Improvements:** Over the years, safety improvements have been made to the circuit, enhancing driver and spectator safety while preserving its challenging character.

- **Modern Era:** Interlagos continues to host the Brazilian Grand Prix, showcasing the best of Formula One in South America. It remains a track that drivers relish for its history and technical challenges.

In summary, the Interlagos Circuit, or Autódromo José Carlos Pace, is a historic and challenging venue in Formula One, known for its undulating layout, unpredictable weather, and passionate Brazilian fans. It has been the setting for many memorable moments in the sport and remains a beloved destination for motorsport enthusiasts worldwide.

Hockenheimring

The Hockenheimring, located in Hockenheim, Germany, is a historic and versatile circuit that has played a significant role in Formula One racing. Known for its rich history and various layout configurations, the Hockenheimring has hosted many memorable races over the years. Here's a brief history behind the Hockenheimring:

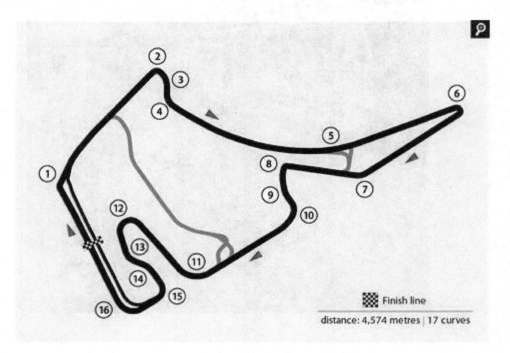

> ➤ **Early Origins:** The Hockenheimring was originally constructed in the early 1930s as a test track for Mercedes-Benz and Auto Union (now part of Audi). It was initially a high-speed circuit with long straights through the forests of the Baden-Württemberg region.

> ➤ **Formula One Debut:** The circuit made its debut on the Formula One calendar in 1970, hosting the German Grand Prix. It quickly gained

recognition for its unique and challenging layout, which combined long straights with tight chicanes and a stadium section.

- **Championship Deciders:** Hockenheim has been the venue for many championship-deciding races, adding drama and excitement to the Formula One season finales. Legendary drivers like Michael Schumacher and Ayrton Senna have clinched titles here.

- **Stadium Section:** One of the circuit's most distinctive features is the "stadium section" towards the end of the lap. This enclosed area is surrounded by grandstands and provides a unique viewing experience for fans.

- **Historic Moments:** Hockenheimring has witnessed historic moments in Formula One, including epic duels and surprise victories. The circuit's layout encourages close racing and overtaking opportunities.

- **Safety Improvements:** Over the years, safety improvements have been made to the circuit, including changes to the track layout to enhance driver safety. The track's high-speed nature led to concerns about safety in the past.

- **Alternating Layouts:** Hockenheimring has alternated between two distinct layouts: the traditional high-speed circuit with long straights and the more modern, compact layout with challenging corners. The choice of layout has varied over different seasons.

- **Fan Engagement:** The Hockenheimring has a dedicated and passionate fan base. German fans, known as "Die Hockenheimring Fans," create a lively atmosphere during race weekends.

- **Modern Era:** Hockenheimring continues to host the German Grand Prix, although the race has not been an annual fixture in recent years due to changing agreements and logistical considerations.

- **Legacy and Heritage:** The Hockenheimring's legacy in Formula One is marked by its historic significance, versatile layout, and ability to deliver exciting races. It remains a circuit with a rich history that has left an indelible mark on the sport.

In summary, the Hockenheimring is a historic and versatile venue in Formula One, known for its alternating layouts, dramatic races, and stadium section that provides a unique experience for fans. It continues to be a part of the Formula One calendar, offering a blend of history and modern racing excitement.

Melbourne Grand Prix Circuit (Albert Park)

The Melbourne Grand Prix Circuit, situated in Albert Park, Melbourne, Australia, is a picturesque and popular circuit on the Formula One calendar. Known for its scenic lakeside setting and vibrant atmosphere, the circuit has been the traditional host of the Australian Grand Prix. Here's a brief history behind the Melbourne Grand Prix Circuit at Albert Park:

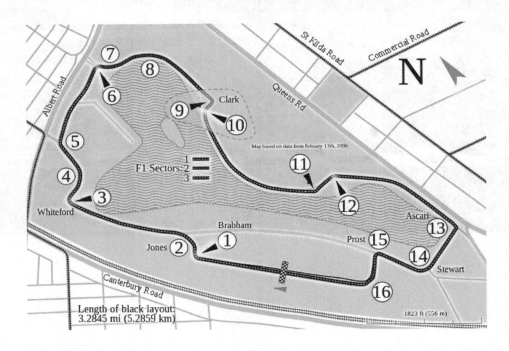

> **Lakefront Beginnings:** The circuit was originally constructed around Albert Park Lake in the 1950s. It was initially used for various motorsport events before becoming the host of the Australian Grand Prix.

- **Australian Grand Prix Venue:** In 1996, the Melbourne Grand Prix Circuit became the official venue for the Australian Grand Prix, replacing the previous host in Adelaide. This move brought the Formula One season opener to Melbourne.

- **Lakeside Scenery:** One of the circuit's standout features is its beautiful lakeside setting. The track winds its way around Albert Park Lake, providing a scenic backdrop for Formula One racing.

- **Temporary Street Circuit:** The Melbourne Circuit is a temporary street circuit, which means that it utilizes existing public roads and parkland pathways for the race. This setup requires the circuit to be assembled and disassembled each year for the Grand Prix.

- **High-Speed Straights and Chicanes:** The layout of the circuit includes high-speed straights followed by tight chicanes and technical sections. This mix of elements makes it a well-rounded and challenging track for drivers.

- **Thrilling Races:** The Melbourne Grand Prix Circuit has witnessed numerous thrilling and unpredictable races, often due to variable weather conditions and the occasional safety car period.

- **Season Opener:** The Australian Grand Prix has traditionally been the season-opening race on the Formula One calendar, adding to the excitement and anticipation surrounding the event.

- **Fan-Friendly Atmosphere:** The Australian Grand Prix is known for its fan-friendly atmosphere, with a variety of entertainment options, fan zones, and activities throughout the race weekend.

- **Modern Era:** The Melbourne Grand Prix Circuit continues to host the Australian Grand Prix, albeit with occasional variations in the race calendar. It remains a favorite destination for both fans and teams.

- **Legacy and Heritage:** The Melbourne Circuit's legacy in Formula One is marked by its picturesque location, exciting racing, and status as the season opener. It has introduced Formula One to a new generation of fans in the southern hemisphere.

In summary, the Melbourne Grand Prix Circuit at Albert Park is a scenic and vibrant venue in Formula One, known for its lakeside setting, thrilling races, and the honor of hosting the season-opening Australian Grand Prix. It combines the beauty of the park with the excitement of top-tier motorsport, creating a unique and enjoyable experience for fans and drivers alike.

Yas Marina Circuit

The Yas Marina Circuit, situated on Yas Island in Abu Dhabi, United Arab Emirates, is a modern and opulent racing venue that has quickly become a prominent fixture on the Formula One calendar. Known for its twilight races, luxurious facilities, and futuristic design, the circuit offers a unique racing experience. Here's a brief history behind the Yas Marina Circuit:

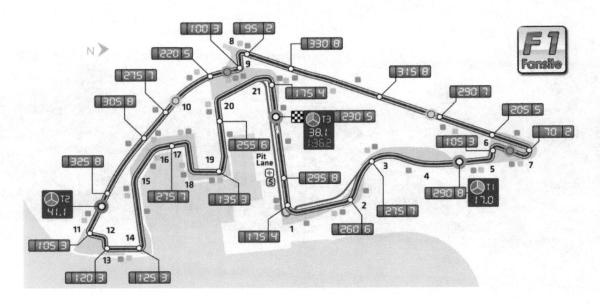

- ➢ **Visionary Development:** The Yas Marina Circuit was conceived as part of the ambitious Yas Island development project, aimed at transforming the island into a world-class leisure and entertainment destination.

- ➢ **Inaugural Race:** The circuit made its Formula One debut in 2009, hosting the inaugural Abu Dhabi Grand Prix. This night race under the floodlights was a groundbreaking addition to the Formula One calendar.

- **Twilight Racing:** The Abu Dhabi Grand Prix is known for its twilight-to-night race transition. The start of the race occurs in the late afternoon, with the sun setting during the race, creating a visually stunning spectacle.

- **Futuristic Design:** The circuit's design is characterized by modern and futuristic architecture, with the Yas Viceroy Abu Dhabi Hotel straddling the track. The illuminated hotel and marina add to the circuit's visual appeal.

- **Pit Lane Exit:** One of the unique features of the Yas Marina Circuit is the pit lane exit that passes under the track, allowing cars to rejoin the circuit safely.

- **State-of-the-Art Facilities:** The Yas Marina Circuit offers state-of-the-art facilities for teams, drivers, and spectators, including luxurious hospitality suites and a marina where fans can arrive by boat.

- **Marquee Event:** The Abu Dhabi Grand Prix is considered a marquee event on the Formula One calendar, attracting not only motorsport enthusiasts but also celebrities and VIP guests from around the world.

- **Iconic Corners:** The circuit features several challenging corners, including the tight and twisty Marina Complex and the long straight leading into the hairpin turn of Turn 7.

- **Entertainment Hub:** Yas Island, where the circuit is located, offers a wide range of entertainment options, including a Ferrari theme park, water parks, and luxurious hotels.

- **Modern Era:** The Yas Marina Circuit continues to host the Abu Dhabi Grand Prix, which often serves as the season-ending race on the Formula One calendar. It brings the curtain down on the season in style.

In summary, the Yas Marina Circuit is a modern and luxurious venue in Formula One, known for its twilight races, opulent facilities, and futuristic design. It has rapidly established itself as a premier destination for motorsport enthusiasts and a symbol of the sport's global expansion into new markets.

Circuit Gilles Villeneuve

The Circuit Gilles Villeneuve, located on Île Notre-Dame in Montreal, Canada, is a historic and beloved circuit in Formula One racing. Known for its challenging layout, scenic surroundings, and passionate Canadian fans, the circuit has hosted the Canadian Grand Prix for many years. Here's a brief history behind the Circuit Gilles Villeneuve:

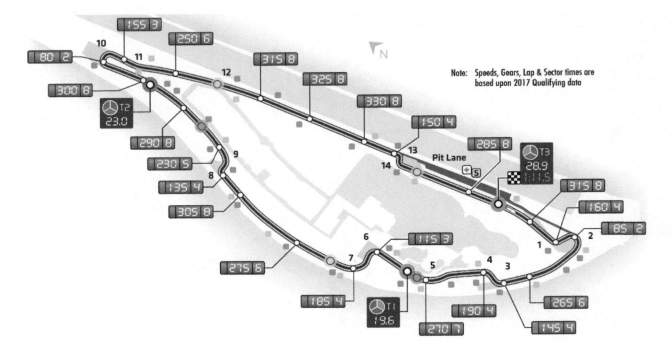

- ➢ **Island Setting:** The circuit is situated on Île Notre-Dame in the St. Lawrence River, creating a picturesque and unique setting for Formula One racing. It was originally built for the 1976 Summer Olympics before being adapted for motorsport.

- ➢ **Inaugural Race:** The circuit made its Formula One debut in 1978, hosting the Canadian Grand Prix. It quickly gained recognition for its challenging layout and memorable races.

- **Named After a Legend:** The circuit is named in honor of Gilles Villeneuve, the legendary Canadian Formula One driver who tragically lost his life during a qualifying session for the Belgian Grand Prix in 1982.

- **Challenging Layout:** Circuit Gilles Villeneuve is known for its challenging and high-speed layout. It features long straights, tight chicanes, and a mix of slow and fast corners, testing the skill and precision of drivers.

- **Wall of Champions:** One of the most famous features of the circuit is the "Wall of Champions." Located at the final chicane, it has earned its name by catching out numerous world champions over the years.

- **Variable Weather:** The Canadian Grand Prix often experiences variable weather conditions, including rain and unpredictable temperatures. This adds an element of unpredictability to the races and can lead to exciting battles.

- **Historic Moments:** The circuit has witnessed many historic moments in Formula One history, including iconic victories, dramatic crashes, and memorable duels.

- **Fan-Friendly Atmosphere:** The Canadian Grand Prix is known for its fan-friendly atmosphere, with passionate Canadian fans, known as "The Eh Team," creating a vibrant and enthusiastic environment.

- **Modern Era:** The Circuit Gilles Villeneuve continues to host the Canadian Grand Prix and remains a favorite among drivers and teams. It is a track that offers a mix of technical challenges and opportunities for overtaking.

- **Legacy and Heritage:** The circuit's legacy in Formula One is marked by its historic significance, challenging layout, and the enduring memory of Gilles Villeneuve, one of the sport's most beloved figures.

In summary, the Circuit Gilles Villeneuve is a historic and challenging venue in Formula One, known for its scenic location, challenging layout, and passionate Canadian fans. It stands as a tribute to the memory of Gilles Villeneuve and continues to be a beloved destination for motorsport enthusiasts worldwide.

Circuit de Barcelona-Catalunya

The Circuit de Barcelona-Catalunya, situated in Montmelo, a suburb of Barcelona, Spain, is a versatile and well-known circuit on the Formula One calendar. Known for its testing significance, rich history, and as a popular pre-season testing venue, the circuit plays a pivotal role in Formula One racing. Here's a brief history behind the Circuit de Barcelona-Catalunya:

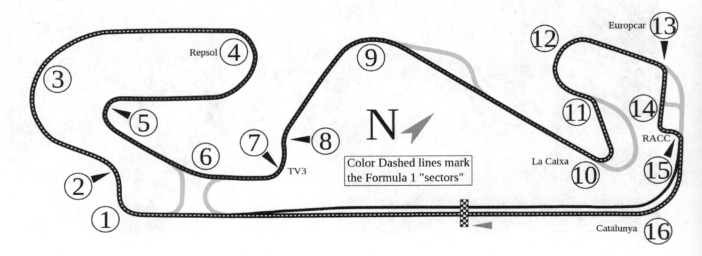

> **Catalan Gem:** The circuit was inaugurated in 1991 and quickly established itself as a top-tier motorsport venue in Spain and internationally. It was designed to host a wide range of motorsport events, including Formula One.

> **Pre-Season Testing:** The Circuit de Barcelona-Catalunya is renowned as a pre-season testing venue for Formula One teams. Its moderate climate and diverse layout make it an ideal location for teams to prepare for the upcoming season.

- **Home to the Spanish Grand Prix:** The circuit has been the traditional host of the Spanish Grand Prix since its opening year in 1991. It showcases the best of Formula One to Spanish fans and international spectators.

- **Varied Layout:** The circuit features a varied layout that challenges teams and drivers with a mix of high-speed straights, medium-speed corners, and technical sections. It is often considered a good barometer of a car's performance due to its versatility.

- **Historic Moments:** Over the years, the Circuit de Barcelona-Catalunya has witnessed many historic moments in Formula One, including intense battles and iconic victories.

- **Spanish Motorsport Hub:** The circuit serves as a hub for Spanish motorsport and offers a wide range of racing events, including MotoGP, FIA World Rallycross Championship, and more.

- **Fan-Friendly Atmosphere:** The Spanish Grand Prix at the Circuit de Barcelona-Catalunya is known for its fan-friendly atmosphere. Spectators can enjoy various entertainment options and activities during race weekends.

- **Modern Era:** The circuit continues to host the Spanish Grand Prix and remains an essential part of the Formula One calendar, both for races and testing.

- **Technical Significance:** Due to its role in pre-season testing, the Circuit de Barcelona-Catalunya often plays a crucial role in shaping the development of Formula One cars for the upcoming season.

- **Legacy and Heritage:** The circuit's legacy in Formula One is marked by its testing importance, versatile layout, and its role in promoting motorsport in Spain and beyond.

In summary, the Circuit de Barcelona-Catalunya is a versatile and important venue in Formula One, known for its pre-season testing significance, diverse layout, and role as the home of the Spanish Grand Prix. It continues to be a hub for motorsport in Spain and a key destination for Formula One teams and fans.

Imola Circuit (Autodromo Internazionale Enzo e Dino Ferrari)

The Imola Circuit, officially known as the Autodromo Internazionale Enzo e Dino Ferrari, is a historic and revered circuit located in Imola, Italy. Known for its challenging layout and storied history in Formula One, the circuit has been a venue for numerous thrilling races. Here's a brief history behind the Imola Circuit:

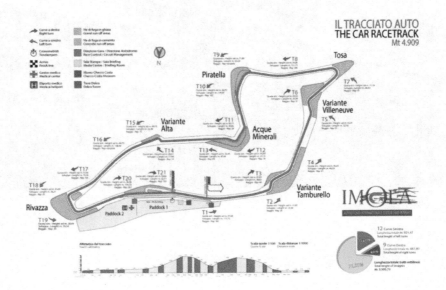

- **Named After Ferrari Family:** The circuit is named in honor of Enzo Ferrari, the legendary founder of the Ferrari automobile company, and his son Dino. It reflects Italy's deep connection to motorsport and its iconic racing heritage.

- **Early Origins:** The Imola Circuit was constructed in the 1950s as a venue for motorsport events. It was initially a high-speed road circuit that used public roads, but over the years, it evolved into a dedicated racing circuit.

➤ **San Marino Grand Prix:** Imola became synonymous with the San Marino Grand Prix, which was introduced to the Formula One calendar in 1981. The race was not held in the independent state of San Marino but at the Imola Circuit due to its central location in Italy.

➤ **Challenging Layout:** The Imola Circuit is renowned for its challenging and technical layout. It features a mix of high-speed straights, tight chicanes, and elevation changes, providing a comprehensive test of a driver's skill and a car's performance.

➤ **Tragic Weekend:** The Imola Circuit is unfortunately best remembered for the tragic weekend in 1994 when two Formula One drivers, Ayrton

Senna and Roland Ratzenberger, lost their lives in separate accidents during the San Marino Grand Prix. These events had a profound impact on motorsport safety and led to significant reforms.

➤ **Safety Improvements:** In the aftermath of the tragic events of 1994, extensive safety improvements were made to the circuit, including modifications to the Tamburello corner where Senna's accident occurred.

➤ **Historic Races:** The circuit has witnessed many historic races, including memorable duels and iconic victories. It continues to hold a special place in the hearts of motorsport enthusiasts.

➤ **Modern Era:** While the San Marino Grand Prix is no longer part of the Formula One calendar, the Imola Circuit returned to the spotlight in 2020 when it hosted the Emilia Romagna Grand Prix. It marked a return to Formula One racing at Imola after a 14-year absence.

➤ **Fan Appeal:** Imola is known for its passionate and knowledgeable fans who create a unique atmosphere during race weekends.

➤ **Legacy and Heritage:** The Imola Circuit's legacy in Formula One is marked by its historic significance, challenging layout, and its role in motorsport safety reforms. It stands as a testament to Italy's enduring passion for motorsport.

In summary, the Imola Circuit, or Autodromo Internazionale Enzo e Dino Ferrari, is a historic and challenging venue in Formula One, known for its storied past, technical layout, and its contribution to motorsport safety. It continues to be a circuit that evokes both respect and emotion from motorsport fans worldwide.

Austria Spielberg Circuit (Red Bull Ring)

The Austria Spielberg Circuit, commonly referred to as the Red Bull Ring due to sponsorship, is a scenic and compact circuit located in Spielberg, Styria, Austria. Known for its breathtaking alpine backdrop, the circuit has gained popularity for its challenging layout and vibrant atmosphere. Here's a brief history behind the Austria Spielberg Circuit, or Red Bull Ring:

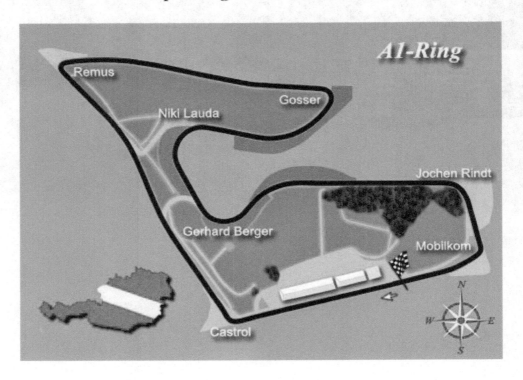

- **Picturesque Setting:** The Red Bull Ring is set against the backdrop of the beautiful Styrian Alps, providing a stunning and picturesque setting for motorsport events.

- **Historic Roots:** The circuit has a rich history, dating back to the 1960s when it was known as the Österreichring. Over the years, it has undergone several transformations and renovations.

- **Ownership by Red Bull:** In 2004, the circuit was purchased by Red Bull founder Dietrich Mateschitz and underwent a major overhaul, leading to its modernization and revitalization.

- **Compact Layout:** The Red Bull Ring features a compact layout with a mix of high-speed straights and challenging corners. It's known for its relatively short lap distance, which results in quick lap times.

- **Formula One Return:** The circuit made its return to the Formula One calendar in 2014, marking the revival of the Austrian Grand Prix. Since then, it has become a fan-favorite destination.

- **Challenging Corners:** The circuit boasts several challenging corners, including the high-speed Turn 9 (Rindt Corner) and the tight Turn 2 (Remus). These corners provide opportunities for overtaking and exciting racing.

- **Fan-Friendly Atmosphere:** The Red Bull Ring is known for its fan-friendly atmosphere, with passionate fans from Austria and neighboring countries attending the races.

- **Scenic Vantage Points:** Spectators can enjoy various scenic vantage points around the circuit, with views of the surrounding mountains and the action on the track.

- **Diverse Racing Events:** In addition to Formula One, the Red Bull Ring hosts various motorsport events, including MotoGP and DTM races, attracting a wide range of motorsport enthusiasts.

- **Modern Era:** The Red Bull Ring continues to host the Austrian Grand Prix and remains a beloved venue on the Formula One calendar, appreciated for its scenic beauty and exciting races.

In summary, the Austria Spielberg Circuit, or Red Bull Ring, is a picturesque and challenging venue in Formula One, known for its compact layout, stunning alpine backdrop, and vibrant atmosphere. It has become a highlight on the racing calendar, offering both drivers and fans an unforgettable experience in the heart of the Austrian Alps.

Chapter 5: Evolution of Engines, Technical Aspects, and Safety in Formula One

1. Engines

1.1 Early Era: In the early years of Formula One, engines were relatively simple, typically naturally aspirated, and featured large displacements. Teams and manufacturers had more flexibility in engine design, leading to a diverse range of power units. This era allowed for innovation and experimentation, giving birth to iconic engines like the V12s and V8s. These engines were characterized by their distinctive sounds and raw power, creating a unique and visceral experience for fans.

1.2 Turbocharged Era: The 1980s marked a significant shift in Formula One as the sport embraced turbocharged engines. These smaller and lighter engines delivered immense power through forced induction, but they also presented technical challenges such as turbo lag and fuel consumption. The turbocharged era was defined by intense competition among engine manufacturers and a relentless pursuit of horsepower. It led to iconic power units like the BMW M12 and the TAG-Porsche engine.

1.3 Hybrid Power Units: In 2014, Formula One entered a new era with the introduction of hybrid power units. These sophisticated power units combine a V6 turbocharged internal combustion engine with energy recovery systems (ERS), including the MGU-H and MGU-K. The goal was to promote fuel efficiency and sustainability while maintaining high-performance standards. This shift reflects Formula One's commitment to align with modern automotive technology and environmental concerns, making the sport more relevant to the wider automotive industry.

1.4 Development Restrictions: To control costs and maintain competitiveness, Formula One introduced strict limitations on engine development. Teams are now limited in the number of engines they can use throughout a season, reducing the expense of constant engine upgrades. This approach aims to level the playing field while encouraging teams to focus on optimizing their power units within the defined parameters, leading to greater reliability and efficiency.

2. Technical Aspects

2.1 Aerodynamics Advancements: Over the decades, aerodynamics became a critical aspect of Formula One car design. Advances in wind tunnel testing, computational fluid dynamics, and CFD simulations led to more complex and efficient aerodynamic solutions. Teams continuously strive to maximize downforce while minimizing drag, resulting in cars that stick to the track and cut through the air with precision.

2.2 Weight and Dimensions: Formula One cars have evolved to become lighter and more compact, with strict weight and dimension limits imposed. These regulations enhance safety by reducing the energy of impacts in accidents. Additionally, lighter cars improve handling and fuel efficiency. Technological advancements in materials and manufacturing have played a crucial role in achieving these weight reductions while maintaining structural integrity.

2.3 Tires: Formula One's tire regulations have evolved to influence race strategy and enhance the excitement of races. The introduction of multiple tire compounds with varying levels of grip and durability has created opportunities for teams to adopt different strategies during races. This has resulted in more unpredictable and thrilling on-track action, as tire management becomes a critical element of racecraft.

2.4 Testing and Development: Formula One has imposed limitations on testing and development to reduce costs and maintain a level playing field among teams. Restrictions on in-season testing and wind tunnel usage have prompted teams to focus on innovation within the confines of these limitations. This approach encourages creativity and efficiency while ensuring that teams do not gain an unfair advantage through extensive testing.

3. Safety

3.1 Safety Regulations: Formula One has consistently prioritized driver safety by introducing comprehensive safety regulations. These regulations cover various aspects, including driver safety gear (such as helmets and suits), car design (cockpit protection), and circuit safety standards (barrier placement, runoff areas). These measures aim to reduce the risk of serious injuries and fatalities in the sport.

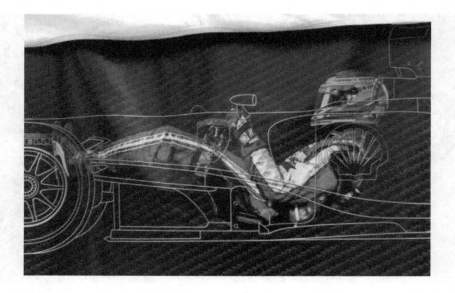

3.2 The Halo Device: In 2018, Formula One introduced the "Halo" cockpit protection device, representing a significant leap forward in driver safety. The Halo device is designed to provide additional protection to a driver's head from potential impacts and flying debris. While initially met with mixed reactions, it has been widely accepted as a crucial safety innovation.

3.3 Medical Car and Procedures: Formula One races are supported by a medical car and a team of medical professionals who are ready to respond to emergencies swiftly. Safety procedures, including the extraction of injured drivers from damaged cars and quick medical assessments, ensure rapid response in case of accidents, minimizing the potential consequences of on-track incidents.

3.4 Circuit Safety: Circuits hosting Formula One races must adhere to strict safety standards. This includes the placement of barriers, runoff areas, and medical facilities strategically positioned around the track. These improvements reduce the risks associated with high-speed racing and provide essential support in the event of accidents.

In summary, the evolution of engines, technical aspects, and safety regulations in Formula One reflects the sport's commitment to innovation, competitiveness, and driver safety. Over the years, Formula One has transitioned from the simplicity of large-displacement engines to the complexity of hybrid power units, embraced advanced aerodynamics, refined tire regulations to enhance race strategy, and implemented rigorous safety measures to protect drivers and spectators alike. Formula One continues to adapt and evolve, remaining at the forefront of motorsport technology and safety standards.

Chapter 5: Drivers' and Constructors' Champions by Year

Drivers' Championship Winners:	Constructors' Championship Winners:
1950: Giuseppe Farina (Italy)	1958: Vanwall (United Kingdom)
1951: Juan Manuel Fangio (Argentina)	1959: Cooper-Climax (United Kingdom)
1952: Alberto Ascari (Italy)	1960: Cooper-Climax (United Kingdom)
1953: Alberto Ascari (Italy)	1961: Ferrari (Italy)
1954: Juan Manuel Fangio (Argentina)	1962: BRM (United Kingdom)
1955: Juan Manuel Fangio (Argentina)	1963: Lotus-Climax (United Kingdom)
1956: Juan Manuel Fangio (Argentina)	1964: Ferrari (Italy)
1957: Juan Manuel Fangio (Argentina)	1965: Lotus-Climax (United Kingdom)
1958: Mike Hawthorn (United Kingdom)	1966: Brabham-Repco (Australia)
1959: Jack Brabham (Australia)	1967: Brabham-Repco (Australia)
1960: Jack Brabham (Australia)	1968: Lotus-Ford (United Kingdom)
1961: Phil Hill (United States)	1969: Matra-Ford (France)
1962: Graham Hill (United Kingdom)	1970: Lotus-Ford (United Kingdom)
1963: Jim Clark (United Kingdom)	1971: Tyrrell-Ford (United Kingdom)
1964: John Surtees (United Kingdom)	1972: Lotus-Ford (United Kingdom)
1965: Jim Clark (United Kingdom)	1973: Lotus-Ford (United Kingdom)
1966: Denny Hulme (New Zealand)	1974: McLaren-Ford (United Kingdom)
1967: Denny Hulme (New Zealand)	1975: Ferrari (Italy)
1968: Graham Hill (United Kingdom)	1976: Ferrari (Italy)
1969: Jackie Stewart (United Kingdom)	1977: Ferrari (Italy)
1970: Jochen Rindt (Austria)	1978: Lotus-Ford (United Kingdom)
1971: Jackie Stewart (United Kingdom)	1979: Ferrari (Italy)
1972: Emerson Fittipaldi (Brazil)	1980: Williams-Ford (United Kingdom)
1973: Jackie Stewart (United Kingdom)	1981: Williams-Ford (United Kingdom)
1974: Emerson Fittipaldi (Brazil)	1982: Ferrari (Italy)
1975: Niki Lauda (Austria)	1983: Ferrari (Italy)
1976: James Hunt (United Kingdom)	1984: McLaren-TAG (United Kingdom)
1977: Niki Lauda (Austria)	1985: McLaren-TAG (United Kingdom)
1978: Mario Andretti (United States)	1986: Williams-Honda (United Kingdom)
1979: Jody Scheckter (South Africa)	1987: Williams-Honda (United Kingdom)
1980: Alan Jones (Australia)	1988: McLaren-Honda (United Kingdom)
1981: Nelson Piquet (Brazil)	1989: McLaren-Honda (United Kingdom)
1982: Keke Rosberg (Finland)	1990: McLaren-Honda (United Kingdom)
1983: Nelson Piquet (Brazil)	1991: McLaren-Honda (United Kingdom)
1984: Niki Lauda (Austria)	1992: Williams-Renault (United Kingdom)
1985: Alain Prost (France)	1993: Williams-Renault (United Kingdom)
1986: Alain Prost (France)	1994: Williams-Renault (United Kingdom)
1987: Nelson Piquet (Brazil)	1995: Benetton-Renault (United Kingdom)
1988: Ayrton Senna (Brazil)	1996: Williams-Renault (United Kingdom)
1989: Alain Prost (France)	1997: Williams-Renault (United Kingdom)
1990: Ayrton Senna (Brazil)	1998: McLaren-Mercedes (United Kingdom)
1991: Ayrton Senna (Brazil)	1999: Ferrari (Italy)
1992: Nigel Mansell (United Kingdom)	2000: Ferrari (Italy)
1993: Alain Prost (France)	2001: Ferrari (Italy)
1994: Michael Schumacher (Germany)	2002: Ferrari (Italy)

1995: Michael Schumacher (Germany)
1996: Damon Hill (United Kingdom)
1997: Jacques Villeneuve (Canada)
1998: Mika Häkkinen (Finland)
1999: Mika Häkkinen (Finland)
2000: Michael Schumacher (Germany)
2001: Michael Schumacher (Germany)
2002: Michael Schumacher (Germany)
2003: Michael Schumacher (Germany)
2004: Michael Schumacher (Germany)
2005: Fernando Alonso (Spain)
2006: Fernando Alonso (Spain)
2007: Kimi Räikkönen (Finland)
2008: Lewis Hamilton (United Kingdom)
2009: Jenson Button (United Kingdom)
2010: Sebastian Vettel (Germany)
2011: Sebastian Vettel (Germany)
2012: Sebastian Vettel (Germany)
2013: Sebastian Vettel (Germany)
2014: Lewis Hamilton (United Kingdom)
2015: Lewis Hamilton (United Kingdom)
2016: Nico Rosberg (Germany)
2017: Lewis Hamilton (United Kingdom)
2018: Lewis Hamilton (United Kingdom)
2019: Lewis Hamilton (United Kingdom)
2020: Lewis Hamilton (United Kingdom)
2021: Max Verstappen (Netherlands)
2022: Max Verstappen (Netherlands)
2023: Max Verstappen (Netherlands

2003: Ferrari (Italy)
2004: Ferrari (Italy)
2005: Renault (France)
2006: Renault (France)
2007: Ferrari (Italy)
2008: Ferrari (Italy)
2009: Brawn-Mercedes (United Kingdom)
2010: Red Bull Racing-Renault (Austria)
2011: Red Bull Racing-Renault (Austria)
2012: Red Bull Racing-Renault (Austria)
2013: Red Bull Racing-Renault (Austria)
2014: Mercedes (Germany)
2015: Mercedes (Germany)
2016: Mercedes (Germany)
2017: Mercedes (Germany)
2018: Mercedes (Germany)
2019: Mercedes (Germany)
2020: Mercedes (Germany)
2021: Mercedes (Germany)
2022: Red Bull Racing RBPT (Austria)
2023: Red Bull Racing RBPT (Austria)

Conclusion: The Timeless Saga of Formula One

As we close the chapter on this comprehensive exploration of Formula One's storied history, we find ourselves standing at the intersection of innovation, bravery, and passion that defines this remarkable sport. Formula One has traversed the decades, evolving into the pinnacle of motorsport, a global phenomenon, and a captivating narrative of human ingenuity, skill, and determination.

A Legacy of Iconic Drivers

Formula One has witnessed the emergence of iconic drivers whose names resonate through the annals of time. From Juan Manuel Fangio's sheer mastery of the sport in its formative years to Ayrton Senna's electrifying speed and intensity, and from Michael Schumacher's relentless pursuit of excellence to Lewis Hamilton's contemporary dominance, these legends have left indelible marks on the sport's history. Each era has brought forth heroes, each with their unique style, charisma, and achievements, captivating fans across the globe.

The Evolution of Speed and Innovation

Formula One cars have evolved from the rudimentary machines of the 1950s to the technologically advanced marvels of today. The sport has witnessed the transition from front-engined monsters to sleek mid-engined rockets, the advent of ground effects and turbochargers, and the introduction of hybrid power units. These innovations have not only pushed the limits of engineering but have also advanced automotive technology, leading to developments that benefit everyday road cars.

The Legendary Constructors

The Formula One paddock has seen iconic teams and constructors rise to glory. Ferrari's passionate tifosi, McLaren's relentless pursuit of perfection, Williams' groundbreaking engineering, and Mercedes' recent dominance have defined different eras. These teams have not only competed for championships but have shaped the sport's history and forged enduring rivalries that have enriched Formula One's narrative.

The Theater of Iconic Circuits

Formula One has graced some of the most iconic and breathtaking circuits on the planet. From the challenging twists and turns of Monaco's streets to the high-speed blasts down Monza's straights, and from the sweeping corners of Suzuka to the modernity of Yas Marina, these circuits have provided the backdrop for historic moments, breathtaking races, and unforgettable duels. The sport's global expansion has brought Formula One to new destinations while preserving the heritage of historic venues.

The Thrill Continues

As we reflect on the history of Formula One, it becomes evident that the thrill, the passion, and the pursuit of excellence are constants that continue to define this remarkable sport. Formula One has adapted to the times, embracing technology, safety innovations, and sustainability while preserving the essence of high-speed racing and competition. It remains a sport where the human spirit meets cutting-edge engineering, where drivers become legends, and where fans from every corner of the world come together to share the thrill of racing.

The story of Formula One is an ongoing saga, and the pages of history continue to be written with each passing season. As we look ahead to the future, the allure of Formula One remains undiminished, promising new chapters of excitement, innovation, and unforgettable moments. Whether you're a lifelong fan or a newcomer to the sport, the world of Formula One welcomes you to join in the journey, as the quest for speed and glory races on.

About the Author

Etienne Psaila, an accomplished author with over two decades of experience, has mastered the art of weaving words across various genres. His journey in the literary world has been marked by a diverse array of publications, demonstrating not only his versatility but also his deep understanding of different thematic landscapes. However, it's in the realm of automotive literature that Etienne truly combines his passions, seamlessly blending his enthusiasm for cars with his innate storytelling abilities.

Specializing in automotive books, Etienne brings to life the world of automobiles through his eloquent prose and an array of stunning, high-quality color photographs. His works are a tribute to the automotive industry, capturing its evolution, technological advancements, and the sheer beauty of vehicles in a manner that is both informative and visually captivating.

A proud alumnus of the University of Malta, Etienne's academic background lays a solid foundation for his meticulous research and factual accuracy. His education has not only enriched his writing but has also fueled his career as a dedicated teacher. In the classroom, just as in his writing, Etienne strives to inspire, inform, and ignite a passion for learning.

As a teacher, Etienne harnesses his experience in writing to engage and educate, bringing the same level of dedication and excellence to his students as he does to his readers. His dual role as an educator and author makes him uniquely positioned to understand and convey complex concepts with clarity and ease, whether in the classroom or through the pages of his books.

Through his literary works, Etienne Psaila continues to leave an indelible mark on the world of automotive literature, captivating car enthusiasts and readers alike with his insightful perspectives and compelling narratives.

Made in the USA
Las Vegas, NV
28 March 2024